LIST OF TITLES

Already published

A Biochemical Approach to Nutrition	R.A. Freedland, S. Briggs
Biochemical Genetics (second edition)	R.A. Woods
Biological Energy Conservation (second edition)	C.W. Jones
Biomechanics	R.McN. Alexander
Brain Biochemistry (second edition)	H.S. Bachelard
Cellular Degenerative Processes	R.T. Dean
Cellular Development	D.R. Garrod
Cellular Recognition	M.F. Greaves
Control of Enzyme Activity	P. Cohen
Cytogenetics of Man and other Animals	A. McDermott
Enzyme Kinetics (second edition)	P.C. Engel
Functions of Biological Membranes	M. Davies
Genetic Engineering: Cloning DNA	D. Glover
Hormone Action	A. Malkinson
Human Evolution	B.A. Wood
Human Genetics	J.H. Edwards
Immunochemistry	M.W. Steward
Insect Biochemistry	H.H. Rees
Isoenzymes	C.C. Rider, C.B. Taylor
Metabolic Regulation	R. Denton, C.I. Pogson
Metals in Biochemistry	P.M. Harrison, R. Hoare
Molecular Virology	T.H. Pennington, D.A. Ritchie
Motility of Living Cells	P. Cappuccinelli
Plant Cytogenetics	D.M. Moore
Polysaccharide Shapes	D.A. Rees
Population Genetics	L.M. Cook
Protein Biosynthesis	A.E. Smith
RNA Biosynthesis	R.H. Burdon
The Selectivity of Drugs	A. Albert
Transport Phenomena in Plants	D.A. Baker

Editors' Foreword

The student of biological science in his final years as an undergraduate and his first years as a graduate is expected to gain some familiarity with current research at the frontiers of his discipline. New research work is published in a perplexing diversity of publications and is inevitably concerned with the minutiae of the subject. The sheer number of research journals and papers also causes confusion and difficulties of assimilation. Review articles usually presuppose a background knowledge of the field and are inevitably rather restricted in scope. There is thus a need for short but authoritative introductions to those areas of modern biological research which are either not dealt with in standard introductory textbooks or are not dealt with in sufficient detail to enable the student to go on from them to read scholarly reviews with profit. This series of books is designed to satisfy this need. The authors have been asked to produce a brief outline of their subject assuming that their readers will have read and remembered much of a standard introductory textbook of biology. This outline then sets out to provide by building on this basis, the conceptual framework within which modern research work is progressing and aims to give the reader an indication of the problems, both conceptual and practical, which must be overcome if progress is to be maintained. We hope that students will go on to read the more detailed reviews and articles to which reference is made with a greater insight and understanding of how they fit into the overall scheme of modern research effort and may thus be helped to choose where to make their own contribution to this effort. These books are guidebooks, not textbooks. Modern research pays scant regard for the academic divisions into which biological teaching and introductory textbooks must, to a certain extent, be divided. We have thus concentrated in this series on providing guides to those areas which fall between, or which involve, several different academic disciplines. It is here that the gap between the textbook and the research paper is widest and where the need for guidance is greatest. In so doing we hope to have extended or supplemented but not supplanted main texts, and to have given students assistance in seeing how modern biological research is progressing, while at the same time providing a foundation for self help in the achievement of successful examination results.

General Editors:

W.J. Brammar, Professor of Biochemistry, University of Leicester, UK

M. Edidin, Professor of Biology, Johns Hopkins University, Baltimore, USA

Muscle Contraction

C.R.Bagshaw

Lecturer
Department of Biochemistry
University of Leicester

Chapman and Hall
London New York

First published 1982
by Chapman and Hall Ltd
11 New Fetter Lane, London, EC4P 4EE
Published in the USA
by Chapman and Hall
733 Third Avenue, New York, NY 10017

Printed in Great Britain by
J.W. Arrowsmith Ltd., Bristol

ISBN 0 412 13450 0

British Library Cataloguing in Publication Data

Bagshaw, C.R.
Muscle contraction.—(Outline studies in biology)
1. Muscle contraction
I. Title II. Series
612'.741 QP321

ISBN 0-412-13450-0

Library of Congress Cataloging in Publication Data

Bagshaw, C.R. (Clive Richard)
Muscle contraction
(Outline studies in biology)
Bibliography: P.
Includes index.
1. Muscle contraction. I. Title. II Series:
Outline studies in biology (Chapman and Hall)
[DNLM: 1. Muscle contraction. WE 500 B149m]
QP321.B26 591.1'852 82-4475
ISBN 0-412-13450-0 AACR2

Contents

Preface

The topic of muscle contraction attracts interest from many disciplines – physiology, biochemistry and biophysics among them. Dialogue between these fields has always been strong but in formulating a molecular mechanism of contraction they have fused. However, the literature has tended to remain scattered between journals and polarized between the level found in general textbooks and that in specialist reviews. My object in writing this book is to provide a bridge to link these sources. No doubt you have already glanced at the contents and have discovered my prejudices. I hope to have struck an acceptable and useful balance, but as a precaution the references given as further reading either extend my treatment or fill in on the background knowledge that I have assumed.

I am indebted to Sir Andrew Huxley, Dr Richard Tregear, Dr David White, Dr Arthur Rowe, Dr Arthur Moir, Dr Michael Geeves, Dr Christine Wells, Kate Poole and Simon Byrne for their comments on one or more sections of the first draft. I am grateful to Dr Hugh Huxley, Drs Michael and Mary Reedy, Dr Roger Craig, Dr Gerald Offer, Paula Flicker and Maria Maw for providing the original photographs used in the half-tone figures. I am grateful to Judith Thompson and the Microcomputing Unit of Leicester University for word-processing.

I am also indebted to the following publishers for their permission to reproduce or base my figures on copyrighted material: W.B. Saunders Co. (Fig. 3.1), Journal of Physiology (Figs 2.6, 6.2, 6.11, 6.12), Rockefeller University Press (Fig. 3.5), Academic Press (London) Ltd (Fig. 4.3) and Longman Group Ltd (Fig. 6.8).

1 Introduction

The ability to move is one of the fundamental characteristics of a living organism, but the mechanisms by which this is achieved are diverse. Muscle contraction is a rather specialized example of this phenomenon, yet the one which is perhaps best understood at the molecular level. Indeed evaluation of its mechanism has helped to advance knowledge of more basic systems of cell motility. This situation contrasts with many other areas of molecular biology where the most detailed information has been obtained by investigating the most primitive organisms.

It is constructive to consider why muscle, the striated variety in particular, remains an appealing system for investigation. Firstly, a large proportion of the cell material is devoted to the contractile function. The two fundamental proteins involved, actin and myosin, comprise 80% of the structural proteins and are therefore available in large amounts for chemical characterization. Secondly, these proteins are arranged in a regular way which provides a clue to their mechanism of interaction. Furthermore these regularities allow the application of diffraction techniques which complement and extend the structural information obtained by microscopy. Thirdly, the contraction occurs on a macroscopic scale. In particular, the unidirectional contraction of a skeletal muscle along its long axis facilitates quantitative measurements of length and tension which can be related to events at the molecular level.

In this book I will focus on the molecular basis of contraction and regulation of the myofibril – the organelle within the muscle cell responsible for movement. Equally important problems from the point of view of the overall function of muscle are its gross anatomical features, its nervous control and the back-up metabolic processes which provide the fuel for contraction. Introductions to these topics are dealt with elsewhere (see Topics for further reading, Chapter 2).

At the molecular level interest extends in at least three directions beyond that of muscle contraction *per se*. Actin and myosin are ubiquitous within eukaryotic cells. These proteins are involved in the movement of cells and the organelles within them. Indeed a striated muscle cell might be viewed as one end of a spectrum in which the myofibrils are relatively permanent structures, whereas in non-specialized cells the contractile components are assembled and dissolved as required. A companion volume in this series, *Motility of Living Cells*, by P. Cappuccinelli, deals with this aspect. Secondly, the actin–myosin system provides a prime example of a biological energy transducer, converting chemical energy to mechanical work. Parallels are therefore sought with other adenosine 5′-triphosphate (ATP)-utilizing enzymes involved in energy transduction, in particular those which give rise to or are driven by concentration gradients. Finally the Ca^{2+} ion has long been associated with the control of contraction. In the last decade there has been a resurgent interest in this chemical messenger, following the discovery of a family of calcium modulated proteins [1]. These proteins are responsible not only for activation of the contractile proteins, but also they coordinate the activation of enzymes involved in providing the fuel supply. In other cells related Ca^{2+}-modulated proteins regulate secretion and division.

Topics for further reading

Offer, G. (1974), The Molecular Basis of Muscular Contraction, in *Companion to Biochemistry*, vol. 1 (Bull, A.T., Lagnado, J.R., Thomas, J.O. and Tipton, K.F., eds), Longman.

Wilkie, D.R. (1976), *Muscle* (2nd edn), Edward Arnold, London.

These introductory texts complement much of the information in this book.

2 Gross anatomy and physiology

2.1 Muscle types

Muscles have evolved to meet a variety of functions which demand gross differences in performance. Skeletal muscles may be required for short bursts of activity or prolonged contractions. Sustained activity is the hallmark of cardiac muscle which can function non-stop for a century or more. The flight muscles of a midge can contract one thousand times a second. A square centimetre of a molluscan adductor muscle can lift a 10 kg weight.

Faced with this diversity, the molecular biologist's priority has been to identify those features which are common, and to focus on the most amenable species for more detailed attention. All muscles appear to involve interaction between actin- and myosin-containing filaments fuelled by ATP hydrolysis. However, the filaments may differ in their arrangement and in the protein isotypes they contain. Muscles also differ in the metabolic reactions they employ to generate ATP and in the way they are controlled by or respond to nerve impulses and chemical effectors. Muscles are classified by any one of these characteristics and this results in an overlapping and confusing array of nomenclature.

Vertebrate striated muscles take on their striped appearance under the light microscope because of the alignment of their myofilaments, whereas vertebrate smooth (unstriated) muscle cells appear almost structureless at this resolution. The latter class include a diverse collection which are generally under involuntary control and are concerned with the slow contraction and constriction of internal organs. Striated muscle may be further divided into skeletal and cardiac which, besides the anatomical layout of their muscle fibres, differ in their excitation mechanism. The former is under conscious control (voluntary), whereas the latter undergoes regular, self-sustained contractions. Individual fibres of a skeletal muscle may respond in an 'all-or-none' fashion to a nervous impulse (twitch) or they may be multi-innervated and produce a graded tonic response. However, a muscle comprising twitch fibres can yield a graded response by changing the number of individual fibres which are activated. The time course of tension development by a twitch fibre may be classified as fast or slow.

Even the colour of the muscle may indicate its function. Red muscle contains a high content of cytochromes and myoglobin, which are proteins associated with oxidative metabolism and allow sustained activity. White muscle relies on glycolysis for rapid ATP synthesis and readily goes into oxygen debt. It is therefore associated with fast-twitch fibres which operate in short bursts. Some muscles may contain a mixture of these fibre types and thus are not suitable for some experimental investigations.

The above classification refers to vertebrate muscles, although invertebrate muscles show some analogous characteristics. They also display some novel ones. Some insect flight muscles oscillate far more rapidly than the frequency of the nervous impulse which stimulates them – hence the coupling is termed asynchronous. The muscle itself only shortens by a few percent during each oscillation and the movement is amplified by a lever system involving the thorax. The wings and the attached ligaments are driven at their natural frequency of oscillation. The nature and arrangement of the filaments of some invertebrate muscles also contribute to their great strength. Their myosin filaments may be bolstered with an additional core protein, paramyosin. Certain molluscan muscles have evolved a mechanism for holding very high tensions with little energy expenditure, the so-called 'catch' state.

For the cases analysed so far, each muscle type from each species appears to have one or more characteristic myosin isoenzymes, differing slightly in amino acid composition and ATPase activity. Ultimately this may allow a molecular classification of muscles, but at present these isoenzymes are named according to their source using the terminology outlined above.

It is useful to consider some specific muscles which are widely studied at the molecular level. In the Introduction the advantages of striated muscle for experimental investigations were outlined. The frog sartorius muscle is favoured for many physiological and structural studies. It is small enough to allow adequate oxygenation and contains fuel reserves for many hundreds of contractions. The fibres within it run parallel to the long axis and are relatively easy to dissect. Unfortunately, it does not provide much scope for biochemical studies, both in terms of quantity and stability of the isolated proteins. For these experiments the usual source is rabbit back (longissimus dorsi) and some leg muscles which comprise largely fast-twitch muscles. Mammalian muscle is not as easy to dissect and does not survive as well as frog muscle. The rabbit psoas muscle is relatively free of connective tissue and has parallel running fibres. It therefore has been widely used for physiological and structural studies. Throughout this book (and most other literature) whenever the muscle source is undefined it should be regarded as being skeletal muscle from one of these species. Extrapolation of the data to other types may not be valid. Comparisons between rabbit and frog muscles show some quantitative differences, but these are often within the variations arising from different experimental conditions. In this regard many of the numbers given in this book are approximate and are used primarily to exemplify the nature of the calculation.

Cardiac and smooth muscles are not easily investigated, but because of their medical importance they have received enormous attention. Bovine hearts provide ample biochemical material, while the papillary muscles within the hearts of smaller mammals are often used in physiological studies because of their shape. The choice of smooth muscle is normally dictated by the question under investigation, but frequently used sources are the walls of the gut and blood vessels.

Invertebrate muscles, besides their inherent interest, provide some practical benefits. Insect flight muscle is comparatively well preserved in electron micrographs and the detail revealed has led to some important concepts in the molecular mechanism of contraction [2]. Some crustacean muscles (e.g. crab and giant barnacle) contain exceptionally large muscle fibres and long sarcomere spacings which aid microinjection work. These have been instrumental in establishing the role of Ca^{2+} in contraction. Molluscan muscles have been investigated in an attempt to elucidate the catch mechanism – the anterior byssus retractor muscle of the mussel being of suitable dimensions for physiological and structural work. However, in characterizing muscles from this phylum a molecular control mechanism was discovered (Section 7.3) which may have a much wider distribution. The scallop striated adductor muscle has been most useful in this respect, yielding a myosin from which the regulatory subunits may be dissociated reversibly [3].

Genetic approaches have not, as yet, had a major impact on the elucidation of the contractile mechanism. The characterization of mutant nematode worms with defective myosins demonstrates the advantages of investigating more primitive organisms [4]. Gene sequences may now be determined more rapidly than protein sequences, although they do not reveal post-translational modification. While it is desirable that a battery of experimental techniques are applied to a single muscle, so that quantitative comparisons are meaningful, it is also important to bear in mind the diversity of these remarkable machines. The choice of the most amenable tissue can outweigh years of technical development.

2.2 Physiological states

Relaxed skeletal muscle is readily extensible. In this condition the actin and myosin filaments do not appear to interact and the elasticity is provided by the associated connective tissue and membranes. On stimulation, the response of the muscle depends on the external constraints. If the muscle is held at a fixed length it will develop tension. Such a process is termed an isometric contraction, even though the muscle does not actually contract (i.e. shorten) in the familiar sense. If the load attached to a muscle is less than the isometric tension the muscle will shorten. The steady velocity of such an isotonic contraction reaches a maximum with zero external load. Applying a force greater than the isometric tension (P_0) will cause a stimulated muscle to extend. Its resistance becomes rather low when the applied load is greater than $2P_0$. A muscle *in vivo* may experience all these conditions during its normal functioning. Note that muscles can only actively shorten. *In situ*, they return to their resting length under the influence of an external force provided by an antagonistic muscle or some elastic structure. Muscles increase their girth on shortening to maintain a constant volume.

If metabolic events are curtailed so that the ATP is not replenished a muscle will become stiff, a state known as rigor. Attempts to stretch a rigor muscle by more than a few percent results in permanent damage to the fibres. Although rigor is a non-physiological condition it has received a great deal of attention because it is a stable (and hence experimentally amenable) state in which the actin and myosin filaments interact strongly. Provided the membrane system is rendered permeable, addition of ATP will cause a rigor muscle to revert to a functional state. It will either contract or relax depending on the Ca^{2+} ion concentration.

2.3 Activation

Skeletal muscle is under nervous control. Let us consider the nature of the signal it receives. The cytoplasm of resting nerve and muscle cells is at a potential of about −60 mV with respect to the bathing medium. This voltage exists because certain ions are prevented from equilibrating across the membrane barrier. Freely permeable ions take up a Nernstian distribution, which represents a balance between their electrical and osmotic energy. In the resting nerve the ratio of external to internal K^+ concentration of 0.1 suggests that this ion is almost at equilibrium. In contrast, Na^+ is maintained at a ratio of 10 by an active sodium pump. On excitation the membrane of a nerve cell becomes temporarily permeable to Na^+, which rapidly enters down its concentration gradient, causing the membrane potential to rise to +40 mV. However, within a few milliseconds K^+ outflow restores the potential to its resting value. This transient depolarization excites the neighbouring area and so the signal (the action potential) is transmitted along the nerve cell in a self-propagating manner. The amount of Na^+ and K^+ actually exchanged, relative to the total content of these ions, is minute because the electrical capacity of a cell is very low.

The nerve terminates in a motor end plate which abuts with the muscle cell. Here the action potential causes the release of a transmitter substance, acetylcholine, which in turn depolarizes the muscle cell membrane. In the case of a twitch fibre the action potential is propagated along its length as described above. The fibre membrane is highly invaginated so that the electrical signal is carried inwards and ultimately stimulates the adjacent vesicles of the sarcoplasmic reticulum to release Ca^{2+} (Section 3.3). The Ca^{2+} concentration in the cytoplasm (sarcoplasm) of the muscle cell rises from about 0.1 μM to 10 μM. Relaxation is achieved by the active reaccumulation of the Ca^{2+} by the sarcoplasmic reticulum. Other muscle types may be activated by different mechanisms. Nevertheless, the ultimate message received by the myofibrillar proteins is a rise in Ca^{2+} concentration and this finding allows the electrical events of the excitation–contraction mechanism to be considered as a separate problem.

An isolated twitch muscle may be stimulated via its attached nerve or,

DISCUSSION

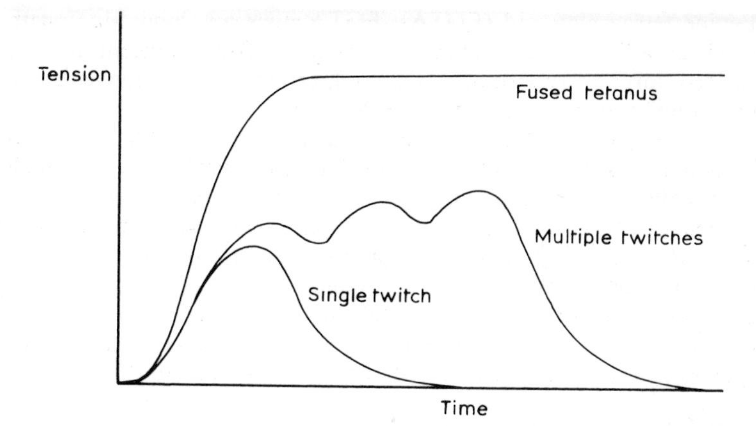

Fig. 2.1 The activation of skeletal muscle by a single and repeated stimulation. A fused tetanus occurs at a stimulation frequency greater than about 10 s^{-1}.

more directly, by applying a small electric shock across its membrane. A characteristic of twitch fibres is that they are maximally activated above a certain threshold stimulus strength – the 'all-or-none' response. In order to activate all the fibres within the muscle it is necessary to apply a supramaximal shock. A single shock causes a twitch contraction. Repeated shocks cause multiple twitches and above a characteristic stimulation frequency these fuse to give a tetanic contraction which remains steady until the stimuli are withheld or until fatigue sets in (Fig. 2.1). Frog skeletal muscle shows an initial lag in tension development of several milliseconds due to the activation mechanism and develops a peak tension in about 50 ms.

If the membrane systems of a muscle cell are damaged the muscle may no longer be electrically excitable. Instead, the concentration of Ca^{2+}, Mg^{2+} and ATP in the vicinity of the myofibril can be controlled directly by the composition of the external bathing medium. Fig. 2.2 shows the relationship between the state of a demembranated muscle and the composition of the medium.

The control system is so sensitive to Ca^{2+} that contaminating amounts in the buffers and glassware are normally sufficient for activation [5]. In practice the Ca^{2+} chelator, ethyleneglycol-bis-(B-aminoethyl ether)-tetraacetic acid (EGTA) is added to reduce the free $[Ca^{2+}]$ to <0.1 μM in order to induce the relaxed state. Unlike ethylene

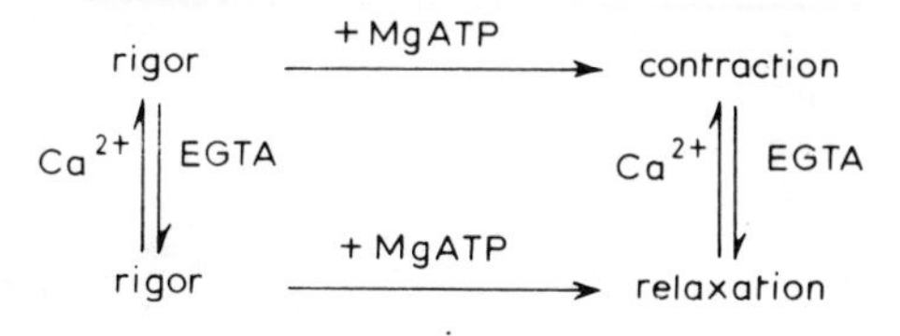

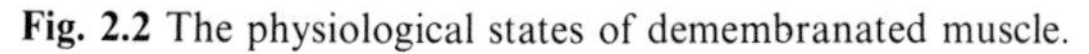
Fig. 2.2 The physiological states of demembranated muscle.

diamine tetraacetic acid (EDTA), EGTA does not bind Mg^{2+} with significant affinity. As we shall see, a high free Mg^{2+} concentration is required for both contraction and relaxation because the effective substrate for the myosin is actually MgATP.

The site of action of Ca^{2+} varies in different muscles. The most well characterized system involves the regulatory protein, troponin, which intrinsically suppresses the interaction between the actin and myosin filaments. Ca^{2+} acts, in effect, by rendering the troponin inoperative. Removal of this regulatory protein therefore results in permanent activation regardless of the Ca^{2+} concentration. This is a useful characteristic because it allows purified mixtures of actin and myosin to be studied, as a model for contraction, independent of the mechanism of regulation. However, not all regulatory systems work in this way.

2.4 Physiological performance

A quantitative assessment of the physiological states alluded to in Section 2.2 remains central to the problem of muscle contraction. A knowledge of muscle performance both defines the problem and provides some clues to feasible answers. Apparatus for measuring the length and tension of a muscle comes in many degrees of sophistication which strive towards increased sensitivity, stability and response time. A simple mechanical device where the muscle operates against a weight illustrates the principles and problems (Fig. 2.3). Such apparatus can be used to define the steady-state contractile properties of muscle but its inertia does not allow much scope for transient analysis. Inertia may be reduced and sensitivity increased by the use of electronic transducers (Section 6.4).

For many muscles the steady velocity of contraction depends on the load as shown in Fig. 2.4. The velocity is at a maximum (V_0) when the muscle is unloaded and declines to zero when the load matches the isometric tension (P_0). Note that a muscle only performs work when its tension, P, operates over a distance. The power output ($P \times V$) reaches a maximum when P and V are about one third their maximum values – a point which cyclists should be aware of when selecting the best gear for an incline. When exerting its isometric tension (P_0) or contracting at its

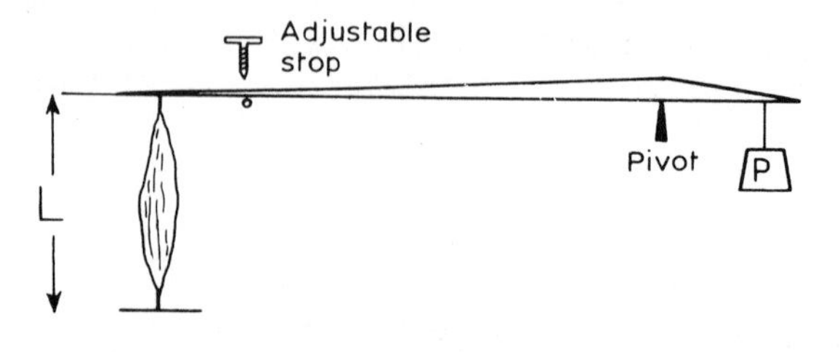

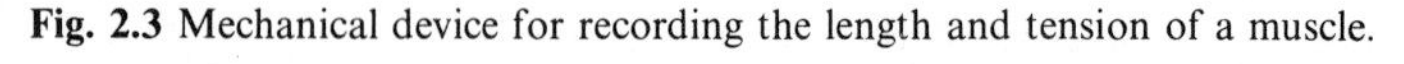
Fig. 2.3 Mechanical device for recording the length and tension of a muscle.

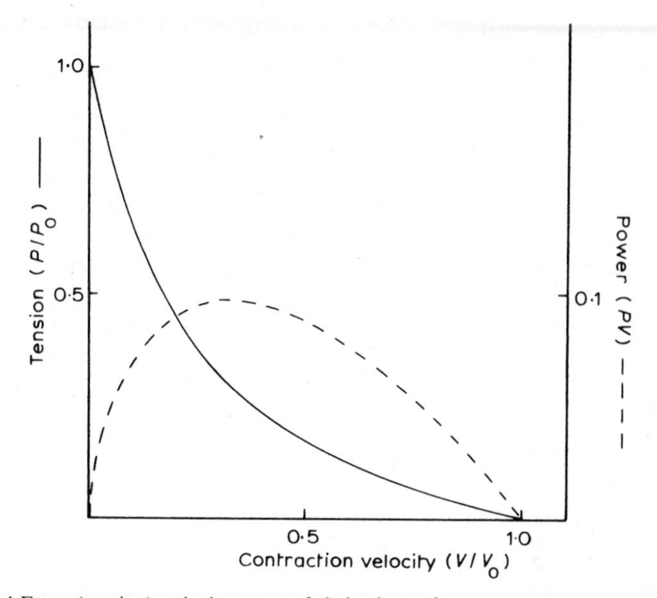

Fig. 2.4 Force (tension)–velocity curve of skeletal muscle.

Table 2.1 Mechanical properties

Term	*Symbol*	*Definition*	*SI units*	*Comments*
Length	L		m	Expressed relative to rest length or as a sarcomere length.
Tension (Force)	T	mass × acceleration	N= kg m s^{-2}	T= internal force which resist applied load (1 kg weight = 9.8 N).
	P	tension/unit area	N m^{-2}	Normalized for cross-sectional area.
Stiffness (elastic modulus)	S	T/L	N m^{-1}	Dimensions depend on whether length and tension are normalized.
		P/L	N m^{-3}	
	E	P/unit length	N m^{-2}	
Compliance	C	L/T	m N^{-1}	Reciprocal of stiffness.
Velocity	V	length/time	m s^{-1}	May be normalized to muscle length/s.
Work	w	force × distance	J(= N m)	May be normalized to sample of unit dimensions.
Power		work/time	W(= J s^{-1})	For constant P and V, power = PV.

maximum velocity (V_0), a muscle does no external work but, as we shall see, it continues to utilize chemical energy. Table 2.1 summarizes the definitions of mechanical terminology, while some illustrative data for a frog sartorius muscle are given in the Appendix.

2.5 Fuel and energetics

The energy source for contraction is derived from the reaction:

$$MgATP^{2-} + H_2O \rightleftharpoons MgADP^{-} + Pi^{2-} + H^{+} \tag{2.1}$$

as demonstrated by the action of MgATP on muscle fibre preparations rendered permeable to the bathing medium. The reaction occurs at sites on the myosin filament and is activated by the actin filament. In intact muscle ATP utilization is not immediately apparent because of the extremely efficient buffer system involving creatine kinase:

$$\text{MgADP} + \text{phosphocreatine (PCr)} \overset{K=100}{\rightleftharpoons} \text{MgATP} + \text{creatine (Cr)} \tag{2.2}$$

Several seconds after the initiation of contraction, glycolysis is activated and ATP is synthesized by substrate level phosphorylation. Phosphocreatine (PCr) is replenished during the recovery process by a net reversal of Equation 2.2 ATP utilization can be demonstrated in contracting intact muscle if creatine kinase is inhibited with fluorodinitrobenzene, glycolysis is blocked with iodoacetate and oxidative phosphorylation is prevented by a nitrogen atmosphere. Under these conditions there is sufficient ATP to support 8 or so twitches whereas an uninhibited excised muscle has a glycogen store sufficient for several hundred contractions.

The ATP molecule (Fig. 2.5) is a source of energy because the equilibrium for its hydrolysis, Equation 2.1, lies far to the right. Several reasons for the increased stability of ADP and Pi (orthophosphate) relative to ATP have been advanced including:

1. The resonance stabilization of the phosphate group. A crude estimate of this effect is obtained by comparing the number of equivalent structures, for the reactants and products, which may be written down using conventional single and double bond notation.
2. At physiological pH, cleavage of the β–γ bond of ATP relieves charge repulsion.
3. The extent of solvation increases on forming the product anions.

Note that the energetics depend on the phosphate moiety *in toto* and

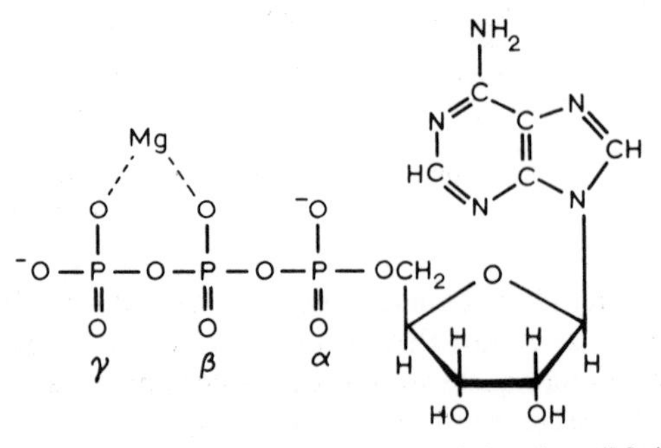

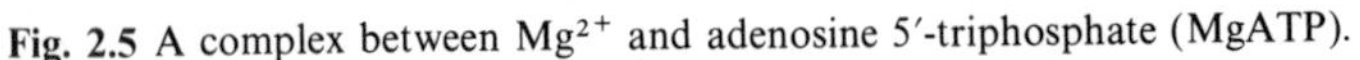
Fig. 2.5 A complex between Mg^{2+} and adenosine 5′-triphosphate (MgATP).

cannot be attributed to a specific high energy bond. The adenosine moiety does not contribute significantly to the energetics, but is important for specific recognition by the ATPase site.

The free energy available from hydrolysis depends on the degree to which the product–reactant concentration ratio, Q, differs from its equilibrium value:

$$\Delta G = RT \ln Q/K \tag{2.3}$$

Frequently the factor $RT \ln(1/K)$ is defined as the standard free energy change, ΔG_0, but care is needed to avoid using this value out of context. ΔG_0 pays no regard to the dimensions of K and invites meaningless comparisons between reactions of differing molecularity.

The evaluation of ΔG for ATP in muscle therefore requires an appropriate *in vitro* determination of K and an *in vivo* determination of the concentrations of all the components involved (ATP, ADP, Pi, H^+, Mg^{2+} at a minimum). K is too large to be determined directly, but it can be calculated from the equilibrium constants of consecutive reactions which result in net ATP hydrolysis, e.g.:

$$\text{ATP} + \text{glucose} \overset{K_1}{\rightleftharpoons} \text{ADP} + \text{glucose-6-phosphate} \overset{K_2}{\rightleftharpoons} \text{glucose} + \text{Pi} \tag{2.4}$$

hence, $K = K_1 . K_2$. The value of K depends on the Mg^{2+} and H^+ concentrations and under conditions appropriate to muscle [ADP][Pi]/[ATP] $= 10^6$ M at equilibrium. In this definition [ADP] includes all relevant species, i.e. $MgADP^-$, $HADP^{2-}$, etc. H_2O is normally omitted from this calculation because its concentration remains practically constant.

The measurement of the actual concentrations of the relevant components in a muscle is far from trivial. Analysis in terms of mol per wet weight of tissue provides some information, but the effective free concentration in the vicinity of the myofibrils may be higher or lower than this value owing to compartmentation or binding. In such an analysis a muscle is clamped between metal tongs, which are precooled in liquid nitrogen, so that reactions are terminated as quickly as possible (within about 80 ms) [6]. The frozen tissue is pulverized and thawed in cold perchloric acid and the denatured protein removed. The remaining metabolites are then analysed quantitatively. A large number of muscles are required for the construction of a time course of metabolite levels during contraction by this method. Complementary techniques are therefore important, in particular those which determine the free rather than total concentrations.

Nuclear magnetic resonance (nmr) spectroscopy has been applied to this problem and much useful information has come from ^{31}P spectra of living muscle [7]. In a strong magnetic field the ^{31}P nucleus (the natural isotope) absorbs radiowaves at a characteristic frequency depending on its chemical environment. PCr, Pi and the three nuclei in ATP give well resolved peaks (Fig. 2.6).

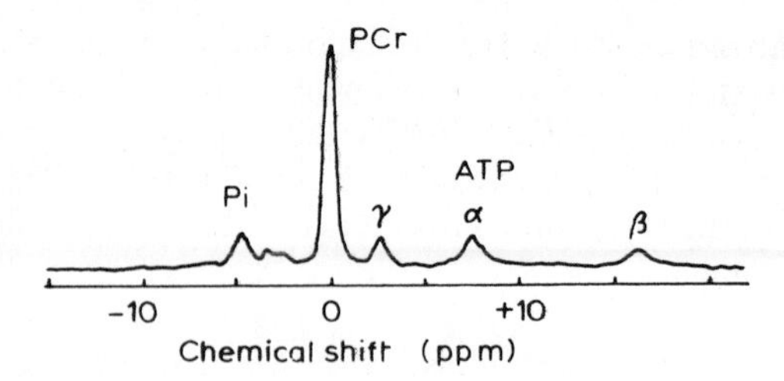

Fig. 2.6 ^{31}P nmr spectrum of frog sartorius muscle. (From Dawson *et al.* [7].)

The advantages of the techniques are:

1. It is non-destructive, hence the same specimen may be examined before, during and after contraction.
2. The area under the spectrum peaks provides a measure of the free metabolite concentrations. Protein bound species are likely to have broadened and shifted peaks and may not be resolved from the background.
3. The exact position of the peaks provides environmental information, e.g. the Pi peak indicates that the sarcoplasm pH is 7.2–7.4 and the ATP peaks suggest this component is present as its Mg^{2+} complex.
4. It may provide information about compartmentation if, for example, Pi is present in pools of differing pH.
5. Fluxes between metabolites can sometimes be evaluated by selectively irradiating one peak and observing the effect on other peaks in the spectrum (saturation transfer). This is appropriate to fluxes on the same time scale as the magnetic recovery process (around 1 s) such as the creatine kinase reaction.

The main disadvantage of nmr is its lack of sensitivity. Several minutes are required to build up sufficient signal to detect components in the mM concentration range. However, a time resolution of about 1 s can be achieved by summing many repeated contractions. When a muscle fatigues, the PCr peak decreases and the Pi peak rises concomitantly and may shift in the acid direction. The ATP peaks remain almost unchanged.

Intracellular pH has also been determined using glass microelectrodes. The measured voltage includes the membrane potential of the fibre (see Section 2.3) which must be recorded with a separate microelectrode and subtracted. Other ion-sensitive electrodes have been developed, including those sensitive to Mg^{2+} and Ca^{2+}, and no doubt these will provide important tools as their sensitivity and selectivity improve. The concentrations of these ions have been estimated by the injection of indicator dyes. This approach has good time resolution, but suffers from interference by other ions and perturbation of the spectra by protein binding.

Table 2.2 Estimates of metabolite concentrations

	ATP	*ADP*	*Pi*	*PCr*	*Cr*	Mg^{2+}
Total (mmol kg^{-1})	4	0.8	2	25	13	10
Free (mM)	4	0.02	2	25	13	3

Table 2.2 provides some estimates of metabolite levels of the sarcoplasm of resting muscle. The degree of confidence is higher where a number of direct methods agree (e.g. ATP). On the other hand, much of the ADP is bound to the actin filament so that the free concentration has to be assessed by calculation, assuming that the creatine kinase reaction remains in equilibrium. It is too dilute to be detected by ^{31}P nmr. From these data (Table 2.2) we may calculate from Equation 2.3 that $\Delta G = -60$ kJ mol^{-1} for ATP hydrolysis within a muscle. This represents the maximum work that can be derived, although in practice a muscle can only trap a proportion of this energy.

The position of the equilibrium of a chemical reaction is determined both by the change in internal energy (enthalpy) and order (entropy):

$$\Delta G = \Delta H - T\Delta S \qquad (2.5)$$

If ATP is hydrolysed in a calorimeter under conditions appropriate to a muscle, the internal energy will be liberated as heat from which ΔH is determined as -48 kJ mol^{-1}. A hypothetical muscle working at 100% efficiency would therefore need to absorb 12 kJ of heat from the surroundings per mole of ATP hydrolysed. In practice muscles do not exceed about 50% efficiency, so that the free energy is dissipated partly as heat and partly as work. However, in all cases, the heat + work output must balance the change in internal energy (ΔH) of the reactions involved, in accordance with the First Law of Thermodynamics. This relationship provides a check on our understanding of the reactions occurring in muscle. The heat + work output of a contracting muscle is largely accounted for by the extent of phosphocreatine breakdown, but the discrepancies are believed to be significant and indicate other heat generating reactions are involved [6].

Even before the phosphate metabolites of muscle were identified, measurements of heat changes demonstrated an important property of the energy transduction mechanism. The rate of heat liberation was found to be higher in a muscle undergoing shortening compared with one developing isometric tension: the so-called Fenn effect. Fig. 2.7a depicts an idealized experiment of this type. Upon stimulation a burst of heat is liberated, associated with the activation mechanism, followed by a steady heat production (curve B). The contribution of the activation heat may be assessed by stimulating a muscle which has been stretched so that the actin and myosin filaments can no longer interact (curve A). If an isometrically contracting muscle is released at time t_1 and allowed to shorten, extra heat is released (curve C). Fig. 2.7b shows the rate of

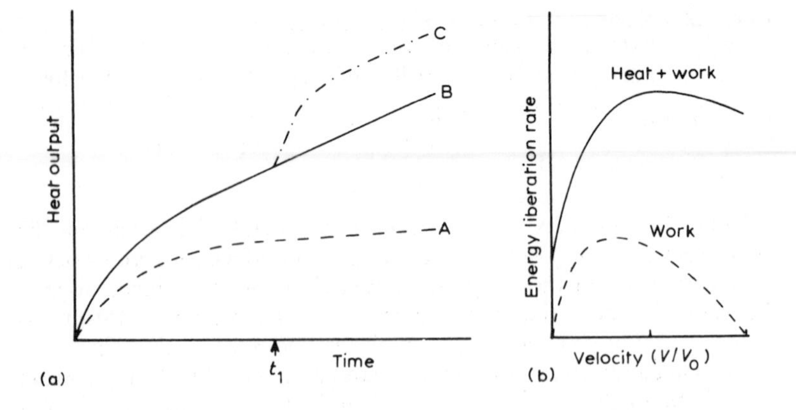

Fig. 2.7 (a) Heat output from a stimulated muscle (A) stretched beyond filament overlap, (B) held isometrically, (C) allowed to shorten at time t_1. (b) Energy liberation rate as a function of contraction velocity.

total energy liberation (heat + work) and power output (work rate) as a function of the isotonic shortening velocity (cf. Fig. 2.4). Freeze clamping experiments confirm that the heat + work output corresponds, to a first approximation, to the phosphocreatine utilization. From such experiments the calculated ATP consumption by frog sartorius muscle at 0°C undergoing isometric contraction is about 500 μmol s^{-1} kg^{-1} muscle, and may rise to 3 times this value on shortening [8]. The ATP turnover by the contractile apparatus of resting muscle is very low and an upper limit of 0.5 μmol s^{-1} kg^{-1} may be calculated from the O_2 consumption [9].

Fig. 2.7b shows that there is a loose coupling between the work output and fuel utilization. Hence a stretched elastic band is not an appropriate analogue of activated muscle, because this liberates a fixed amount of heat + work regardless of the speed of shortening. Rather, a muscle behaves like an automobile where fuel consumption is related to speed and load. The analogy may be taken further. An isometrically contracting muscle hydrolyses considerably more ATP than a resting muscle, but the energy is released as heat. Likewise a motor car may be held on an incline at the expense of fuel and a hot clutch, but no work is performed. A 'hand-brake' mechanism is found in the catch muscles of molluscs, but vertebrate muscles can only develop a high tension at zero cost in the non-physiological state of rigor. At high velocities of contraction the ATP consumption declines, suggesting that the number of effective actin–myosin interaction sites falls. Car manufacturers have also recognized that cruising at high speeds does not demand as much fuel as acceleration and have devised fuel injection systems which automatically reduce the number of cylinders which are fed.

This characteristic performance suggests and limits plausible mechanisms of contraction and should be borne in mind in subsequent chapters which focus on the components involved. Many procedures for preparing these components might be compared with stripping an engine with a sledge hammer. Even if the vital parts remain intact we still need a guide as to how they may be reassembled in a functional way.

Topics for further reading

Aidley, D.J. (1978), *The Physiology of Excitable Cells* (2nd edn), Cambridge University Press. (Provides a good introduction to excitation–contraction coupling.)

Alexander, R. McNeill (1975), *Biomechanics*, Chapman and Hall, London. (Includes a description of muscle function at the macroscopic level.)

Carlson, F.D. and Wilkie, D.R. (1974), *Muscle Physiology*, Prentice-Hall, New Jersey. (Covers the performance and metabolism of muscles.)

White, D.C.S. (1974), *Biological Physics*, Chapman and Hall, London. (A refresher course in physics relevant to muscle and techniques.)

3 Muscle cells

3.1 Striated muscle

A schematic dissection of a skeletal muscle is shown in Fig. 3.1. At low power magnification muscle is seen to comprise bundles of fibres sheathed in connective tissue. Each fibre is a giant multinucleated cell, formed by the fusion of many myoblast cells during its development. Individual fibres vary greatly in size from a few mm to many cm in length and 20–100 μm in diameter, and in some cases they may span the entire length of the muscle. Within each fibre, and occupying about 80% of its volume, are a thousand or so myofibrils, the rod-like organelles responsible for contraction. Mitochondria are sandwiched between the myofibrils and the nuclei are forced to the periphery.

A striking feature of the fibre is the presence of cross bands arising from aligned striations on each myofibril. Although myofibrils are transparent, these bands were seen by early microscopists by slightly underfocusing, and they attributed this effect to areas of high refractive index (i.e. high protein concentration). Resolution of these bands by refraction rather than by staining is important because it allows living muscle to be observed. A number of microscopes have been developed for this purpose. Light passing through a region of high refractive index is slowed down and hence its phase is shifted relative to an uninterrupted beam. Phase contrast microscopy detects such a region by accentuating the interference between the diffracted light, which passes through the region, with the light which passes around its edge. In interference microscopy the illuminating beam is actually split by a half silvered mirror and is recombined after the passage of one beam through the specimen. This technique therefore allows a quantitative estimate of the refractive index throughout the specimen. If the elements (i.e. proteins) within a region responsible for its refraction are not distributed homogeneously, then the refractive index will depend on the plane of polarization of the light (birefringence). Such is the case with striated muscle. Under the polarizing microscope the protein dense bands are seen to be anisotropic with respect to refractive index, while the less dense regions between them are relatively isotropic, hence the terminology A- and I-bands. In the centre of the I-band a highly refractive Z-line is apparent, while the middle of the A-band is less dense, giving the H-zone.

The full significance of this banding pattern is revealed by the additional resolution of electron microscopy (Fig. 3.2). Here it can be seen that the bands arise from the interdigitation of sets of filaments.

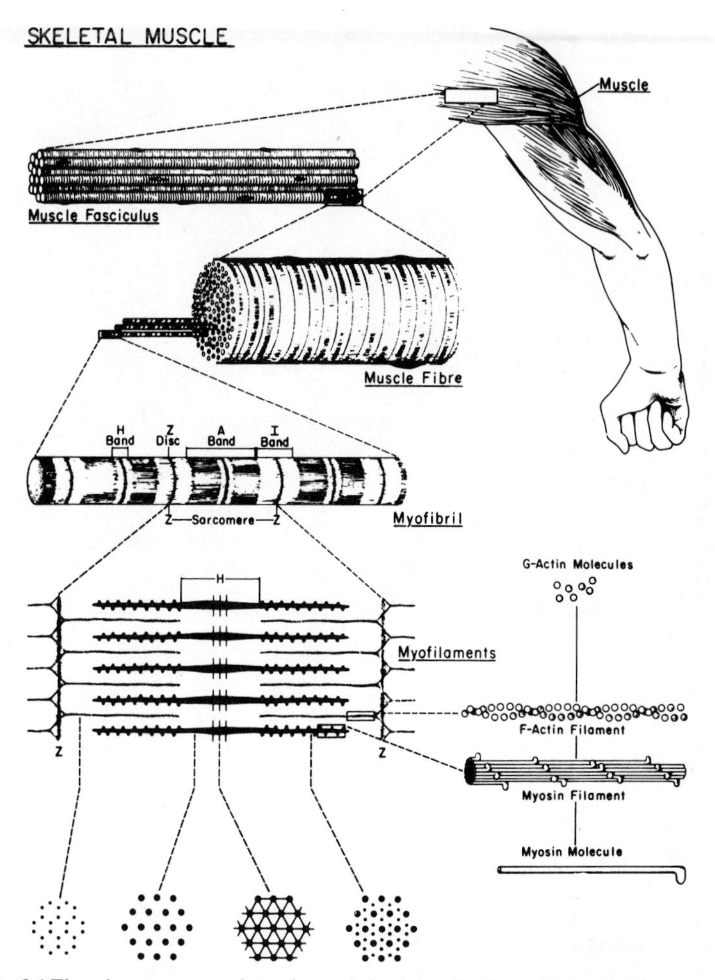

Fig. 3.1 The micro-anatomy of vertebrate skeletal muscle. (From Bloom and Fawcett, see Topics for further reading.)

Thin filaments emanate from the Z-line and make up the I-band, while thick filaments make up the A-band. The thin filaments extend into the A-band and so enhance its density, leaving the paler H-zone in the non-overlap region. In addition, an M-line is revealed in the centre of the H-zone which holds the thick filaments in register. The unit between two Z-lines is defined as a sarcomere. In cross-section the filaments are arranged hexagonally with one thick filament surrounded by six thin filaments. The filament spacing varies from about 20 to 30 nm depending on the sarcomere length.

When a muscle fibre is illuminated by a narrow beam of monochromatic light the striations give rise to an optical diffraction pattern. The separation of the fringes allows calculation of the average sarcomere

Fig. 3.2 Electron micrograph of rabbit psoas muscle. (Courtesy of Maria Maw.)

spacing. While this is a convenient, non-destructive method, care is needed in critical studies because local areas of disorder may go undetected.

3.2 The sliding filament theory

Using interference microscopy to view living muscle fibres, A.F. Huxley and Niedergerke [10] showed that, on stretching or shortening, the A-bands remain at constant length while the I-bands change. Phase contrast examination by H.E. Huxley and Hanson [11] revealed that when isolated myofibrils were induced to contract by addition of ATP, the I-band and H-zone shortened in unison. These findings, taken in conjunction with electron micrographs which demonstrated the underlying basis for the striations, led to the sliding filament theory in which contraction was proposed to occur solely by the interdigitation of the thick and thin filaments. The sarcomere length depends on the degree of overlap while the filaments themselves remain at constant length. In frog skeletal muscle the thick filaments, and hence A-bands, are 1.6 μm long and the thin filaments extend 1.0 μm either side of the Z-line. At rest length, the sarcomere spacing is about 2.6 μm long and hence the overlap is 0.5 μm in each half-sarcomere.

The thick filament comprises mainly myosin and the thin filament, actin; as shown by selective extraction of the A- and I-bands. At high magnification, electron micrographs of muscle in rigor reveal connections linking the thick and thin filaments in the overlap zone: the so-called crossbridges (Fig. 3.3). In muscles fixed under relaxing conditions

Fig. 3.3 Electron micrograph of insect flight muscle in rigor showing crossbridges linking thick and thin filaments. (Courtesy of Drs M. and M.K. Reedy.)

the crossbridges are not so clearly defined, but they appear to emanate from the thick filament. Examination of individual filaments, obtained by the mechanical disruption of relaxed muscle, confirms the source of the crossbridge. Projections are apparent along the length of the isolated thick filament (apart from a central bare zone). In some cases, the crossbridge arrangement is sufficiently well preserved to give rise to regular cross-striations (Fig. 3.4). As we shall see, the crossbridge is an integral part of the myosin molecule and its structure and function hold the key to the mechanism of contraction.

3.3 Membrane systems

The membrane which surrounds a muscle fibre, the sarcolemma, periodically invaginates the fibre to form T tubules. Along each side of a T tubule lie the internal membrane compartments of the sarcoplasmic reticulum, so that in a longitudinal section of muscle a triad of vesicles is seen (Fig. 3.5). As alluded to in Section 2.3 the function of the T tubule is to transmit the action potential into the region of the contractile machinery where it stimulates the sarcoplasmic reticulum to release Ca^{2+}. In this way the diffusion time for the Ca^{2+} message is minimized.

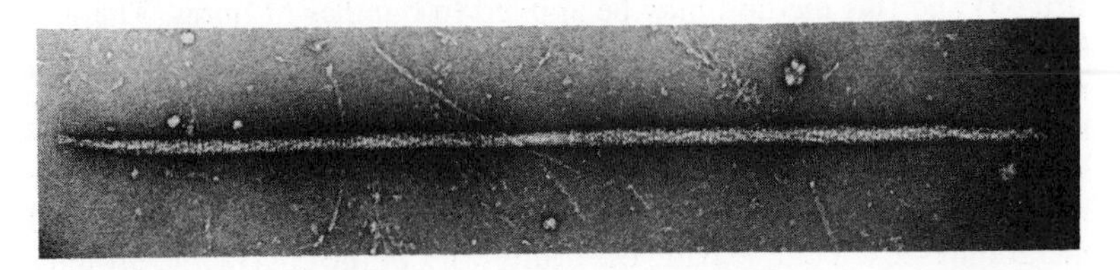

Fig. 3.4 Electron micrograph of a scallop adductor muscle thick filament. The faint cross-striations arise from the regular disposition of the crossbridges at 14.5 nm spacing. (Courtesy of Dr R. Craig.)

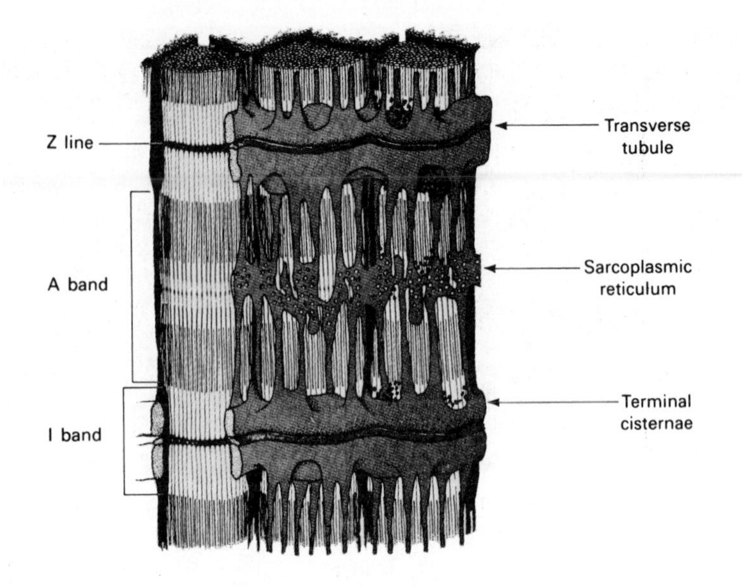

Fig. 3.5 The sarcoplasmic reticulum of frog skeletal muscle. (From Peachey [12].)

Triads appear at each Z-line in frog sartorius muscle, but in other muscles they may appear at the boundary of the A- and I-bands [12].

3.4 Fibre preparations

The membrane system provides a barrier to the bathing medium and prevents experimental manipulation of the chemical environment in the vicinity of the myofibrils. There are several methods for overcoming this problem.

Individual fibres may be mechanically skinned by peeling of the sarcolemma. The key to successful skinning is that the bathing medium must be free of Ca^{2+} ions, otherwise the fibre will be activated and develop local clots of contracted material. Once the operation is complete, however, the skinned fibre can be induced to develop a steady tension by raising the free Ca^{2+} level. The sarcoplasmic reticulum remains intact but its capacity to accumulate Ca^{2+} is swamped by the bathing medium.

Skinning may also be performed chemically by using detergents (e.g. Brij 35) and this method may be applied to bundles of fibres. The first successful permeable muscle preparation was made by immersing a muscle in 50% aqueous glycerol which, by a combination of osmotic shock and solubilization, disrupts all the membrane components. The solution also acts as an antifreeze so that the muscle can be stored for many months at $-20°C$. During this time the soluble proteins and metabolites leach out, leaving the framework of the contractile proteins in a rigor state.

Although it is clear that ATP causes these preparations to contract, care is needed in quantitative work to ensure that diffusion does not become a rate limiting factor. To avoid this the fibre may be preincubated in a solution containing creatine kinase and phosphocreatine so as to provide an ATP regenerating system within the myofibrils.

Fragments of myofibrils can be prepared by mechanically disrupting a muscle with a blender, and these may be purified by washing at low ionic strength. To prevent precontraction the muscle is allowed to go into rigor with its ends fixed before blending. Myofibril preparations are convenient for microscopy and biochemical studies but they suffer from a severe limitation that, in general, their ends cannot be fixed. After a one-off shortening their structural order is irretrievably lost. In their classic study, Huxley and Hanson [11] took advantage of myofibrils whose opposite ends had fortuitously adhered to the slide and coverslip respectively, so enabling them to adjust the sarcomere length.

3.5 Unstriated muscle

The striations of skeletal muscle reflect the alignment of the myofilaments within and provide direct evidence for a sliding filament mechanism of contraction. Is this intricate order fundamental to the mechanism of sliding and hence is this mechanism restricted to striated muscle? Studies on synthetic actomyosin threads, made by extruding solubilized actin and myosin into low ionic strength solution, suggest not. On addition of ATP the thread shortens, yet the filaments within it are disorganized. Indeed when first observed in the 1940s, this phenomenon was taken as evidence that contraction involved gross changes in the lengths of the protein molecules themselves. In the light of the sliding filament mechanism it appears that net shortening of the thread occurs, despite the rather random orientation of the filaments, because those favourably aligned will contribute to a local contraction and buckling. These observations suggest that the regular arrangement of filaments in striated muscle has evolved to allow rapid and efficient contraction, but this order is not fundamental for movement.

Smooth muscle comprises spindle-shaped, mononucleate cells with no obvious cross-striations. In the electron microscope longitudinally-running actin filaments are resolved, but the state of the myosin remains in question. *In vitro*, at least, the myosin forms filaments. A similar situation may exist in non-muscle cells where actin and myosin are involved in shape changes and movement of organelles. In these primitive contractile systems it appears that the actin and myosin filaments are assembled as required. Motion might then be achieved by filament sliding.

Topics for further reading

Bloom, W. and Fawcett, D.W. (1975), *A Textbook of Histology* (2nd edn), W.B. Saunders Co., Philadelphia. (Contains a good selection of micrographs of muscle.)

Huxley, H.E. (1973), Muscle Contraction and Cell Motility, *Nature*, **243**, 445–449. (A review lecture.)

4 Protein components

4.1 Proteins of the myofibril

Further dissection of the contractile machinery goes beyond the resolution of mechanical manipulation and brings us to the biochemical approach of solubilization. This is usually achieved with high ionic strength salt solutions. Once solubilized, the protein components can be purified and characterized. Evidence for the location of a specific protein in the myofibril may come from four lines:

1. Selective extraction may be revealed by microscopy of the remaining myofibril.
2. The protein itself may form a structure recognizable by electron microscopy.
3. A protein may copurify with a major component whose location is known.
4. Specific antibodies may be raised against the protein by injecting it into a different species. These are purified from the host's serum and added to a myofibril preparation of the first species where they bind to the protein under question. The antibody may be apparent from its own mass in electron micrographs, or it can be labelled with a fluorescent reagent and examined by optical microscopy.

Myosin is selectively extracted from a mince of rabbit skeletal muscle with 0.3 M KCl, 0.15 M KPi at pH 6.5. It is recovered by precipitation at low ionic strength and purified by repeated solubilization. Actin is extracted from the residue of the mince. Selective extraction is not successful with all muscles. A more general preparative procedure involves dissolving washed myofibrils in 0.5 M KCl to give actomyosin (a mixture of actin and myosin) and separating the two components by ultracentrifugation or ammonium sulphate precipitation in the presence of ATP.

Preparations are characterized by electrophoresis on a polyacrylamide gel in the presence of the detergent, sodium dodecyl sulphate (SDS) (Fig. 4.1). SDS dissociates proteins into their subunits and binds to the resultant polypeptide chains to give them a similar negative charge per unit length. As the polypeptides permeate the gel matrix towards the anode they separate according to their molecular size. The stoichiometry of the components may be estimated from the intensity of staining, but problems can arise from unresolved bands, anomalous stain uptake and adventitious proteolysis.

Myosin (55%) and actin (25%) are the major components of the

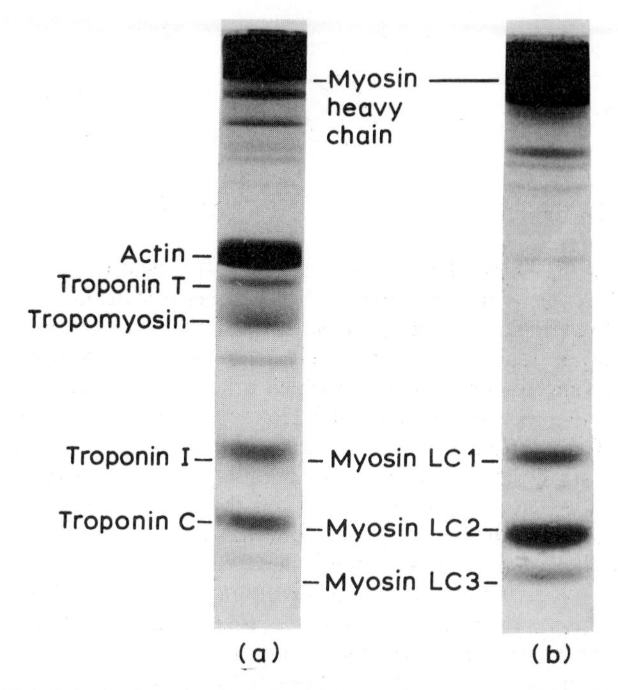

Fig. 4.1 Gel electrophoresis of muscle proteins in the presence of SDS. (a) Solubilized rabbit skeletal muscle myofibrils. (b) Rabbit fast skeletal myosin.

myofibril. The regulatory proteins troponin and tropomyosin are associated with the actin (thin) filament. The myosin (thick) filament contains a number of minor components, among them M-line protein and C protein which are possibly involved in filament assembly. Invertebrate thick filaments contain a core of paramyosin and in some muscles this may be the dominant component. α-Actinin is found in the Z-line.

4.2 Actin

Actin is extracted from an acetone-dried powder of minced muscle by low ionic strength solutions. On adding salt the actin solution becomes viscous as a result of the polymerization of globular G-actin to fibrous F-actin (Fig. 4.2).

G-actin comprises a single polypeptide chain (mol. wt. 42 000), of known sequence, which binds Mg^{2+} and ATP reversibly. On polymerization the ATP is hydrolysed to give one tightly bound ADP molecule per subunit, but this reaction is not obligatory for the formation of F-actin. F-actin filaments have a polarity because of the asymmetry of the G-actin subunits. *In vitro* at physiological ionic strength, F-actin exists in dynamic equilibrium with a low critical concentration (~ 1 μM) of G-actin, crudely analogous to the situation found with lipid micelles. Actin is ubiquitous in eukaryotic cells where it can elicit shape changes by two

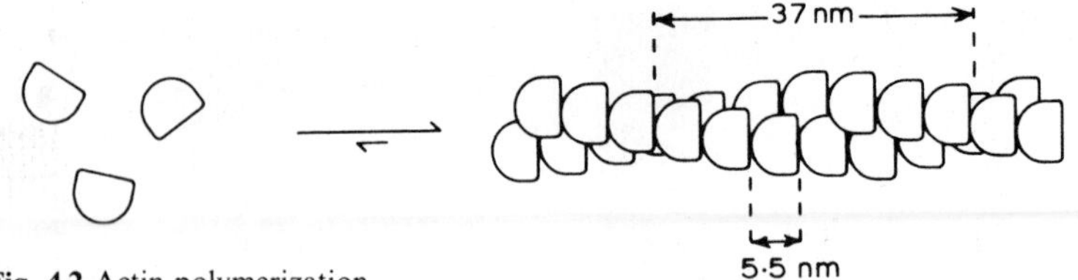

Fig. 4.2 Actin polymerization.

different mechanisms. The gel–sol state of the cytoplasm may be controlled by the interconversion of F- and G-actin. Ancillary proteins regulate this transformation by blocking the polymerization or depolymerization reaction. Alternatively, the F-state may provide a scaffolding (microfilaments), along which other proteins (e.g. myosin) can interact in a vectorial manner. Striated muscle represents an extreme example of the latter, in which the F-actin of the thin filament is a permanent structure of constant and precisely determined length. F-actin activates the ATPase of myosin and this reaction is the driving force behind filament sliding. Apart from this function, actin is usually assumed to have a rather passive mechanical role in muscle contraction. There are no grounds for this assumption; however, actins from different muscle types are much more conserved in sequence and properties than are the corresponding myosins.

The enzyme, DNase I, forms a tight complex with G-actin and is a potent depolymerizing agent. The complex has been crystallized and its structure determined by X-ray diffraction [13]. It is apparent that the G-actin moiety is bi-lobed rather than spherical. The double-stranded helical arrangement of the G-actin units in the thin filaments has been elucidated from electron micrographs and X-ray diffraction of actin paracrystals and intact muscle (Section 6.3).

4.3 Myosin

Myosin is an intriguing molecule. It self-associates to form thick filaments, it hydrolyses ATP and interacts with actin to produce motion. Furthermore, these interactions may be controlled by the Ca^{2+} ion, directly or indirectly, via regulatory subunits which are an integral part of the molecule. Under denaturing conditions myosin (mol. wt. 480 000) dissociates into two heavy chains (mol. wt. 200 000) and four light chains (mol. wt. 20 000). Early hydrodynamic studies showed that myosin was highly asymmetric and in the electron microscope it is seen to consist of two pear-shaped heads attached to a long tail (Fig. 4.3). Each heavy chain forms the bulk of one head and intertwines with its neighbour to form the tail (Fig. 4.4). The amino acid sequence is largely known for rabbit skeletal and nematode body wall myosins [4]. The N-terminus is in the head region.

The structure–function relationship of such a large molecule has been unravelled by examining its proteolytic fragments. Myosin is particularly susceptible to proteolysis at two locations in each heavy chain

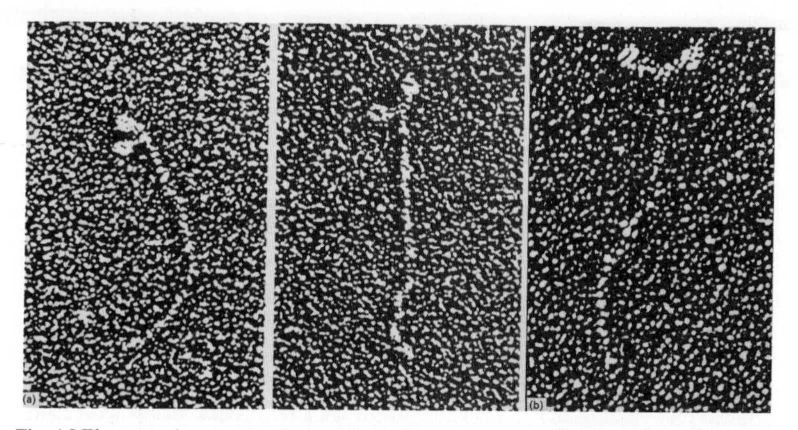

Fig. 4.3 Electron micrographs of (a) rabbit skeletal myosin. (Courtesy of Drs A. Elliott and G. Offer [14].) (b) Scallop adductor myosin. (Courtesy of Paula Flicker.)

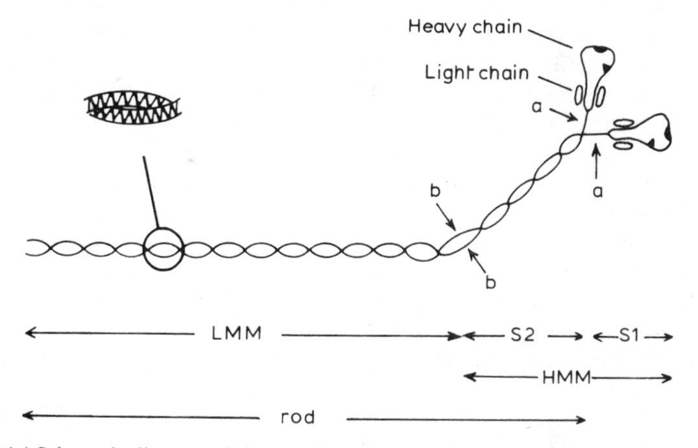

Fig. 4.4 Schematic diagram of the myosin molecule. Proteolytic enzymes attack at the points indicated (papain (a); chymotrypsin (a and b); trypsin (b); to release the fragments shown. The actin and ATP binding sites are located on the heads. The inset indicates the coiled-coil structure of the tail.

(Fig. 4.4) and cleavage at one or both sites depends on the enzymes and conditions used [15]. Some properties of the resultant fragments are given in Table 4.1. These fragments account for the bulk of the parent molecule, but short peptides may be released from the susceptible regions and the light chains may be degraded.

The limited digestion products show that the ATP, actin and two light chain binding sites are located on each subfragment 1 (S1) moiety and that light meromyosin (LMM) accounts for the self-association of myosin at low ionic strength. X-ray diffraction, optical rotatory dispersion and sequence analysis show that the LMM portion forms a coiled-coil of α-helix. Every alternate 3rd and 4th residue is

Table 4.1

Protein	*Symbol*	*Mol. wt.*	*Length (nm)*	*Self association*	*Actin and ATP sites*
Myosin	M	480 000	160	+	2
Light meromyosin	LMM	140 000	100	+	0
Heavy meromyosin	HMM	340 000	60	−	2
Subfragment 1	S1	120 000	15	−	1
Subfragment 2	S2	60 000	40	−	0
Rod		200 000	140	+	0

hydrophobic, while between them are clusters of charged residues. In each α-helix (3.5 residues per turn) the hydrophobic residues form a continuous face which is responsible for binding to the partner heavy chain. The clusters of charged residues on the outside of the coil cause LMM to self-aggregate at low ionic strength. Myosin, itself, forms aggregates which resemble native thick filaments in having projections along their length and a central bare zone (cf. Fig. 3.4). The projections are lacking in aggregates of LMM and rods, and have therefore been identified as the myosin heads. The presence of a bare zone demonstrates that synthetic myosin filaments are bipolar, but they tend to be shorter and wider than native thick filaments. Although the general arrangement of the myosin molecules within the thick filament is known (Fig. 4.5) the precise details vary in different species and remain in question (Section 6.3).

Subfragment 2 (S2) has a sequence pattern similar to LMM, but it forms a coiled-coil which is soluble at physiological ionic strength. Hence it allows the S1 to project a variable distance (up to 55 nm) from the filament backbone. In the absence of crystals suitable for X-ray diffraction, the detailed structure of the S1 head remains unknown.

The light chain subunits are located on the S1 moiety, probably near to the S1–S2 junction. Each head contains two light chains which differ in structure and properties, but insufficient is known about them to allow a completely rational nomenclature. Neither type is required for ATPase activity [16], but in some species it is clear that they regulate or modulate the ATPase in the presence of actin (Section 7.3). Three light chain bands are observed in fast skeletal muscle myosin preparations because one class exists in two isotypic forms (Fig. 4.1).

Considerable effort has been directed towards identifying conformational changes in myosin. Addition of ATP causes no detectable change in the amount of secondary structure as assessed by circular

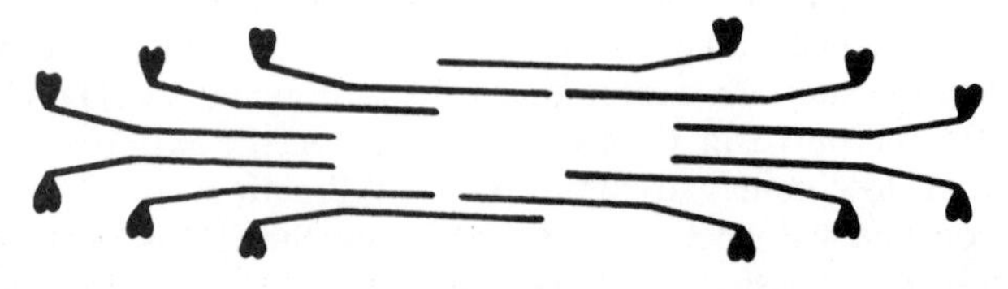

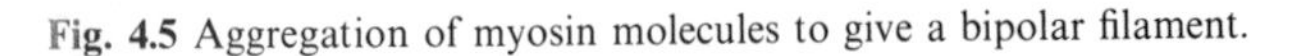

Fig. 4.5 Aggregation of myosin molecules to give a bipolar filament.

dichroism. Changes are observed in the tryoptophan fluorescence but these could involve small shifts typical of most enzymes whose active site is induced to fit around the substrate. A significant movement within the myosin head during ATPase activity involves two reactive cysteine residues, SH-1 and SH-2 [17]. These residues are not part of the catalytic site, although labelling them does affect the ATPase activity. SH-1 and SH-2 are separated by 9 residues in the sequence and by 1 nm in space as judged by cross-linking reagents. However, in the presence of ADP and oxidizing agents, SH-1 and SH-2 react directly to form a disulphide bond and trap ADP at the active site.

One important question which has been tackled concerns the degree of flexibility of the heads about the tail. The fact that proteolytic enzymes of differing specificity can attack the same region suggests that the LMM–HMM and S1–S2 junctions are open and flexible. Furthermore, electron micrographs of myosin preparations show a high proportion of molecules with bends at various angles at these points [14]. For a more quantitative assessment spectroscopic techniques have been applied which report on the orientation change of a probe (containing a dipole) rigidly attached to the myosin head. The fluorescent reagent iodoacetylsulphonaphthyl ethylene diamine (IAEDANS) labels the reactive cysteine residue (SH-1) in S1 and takes on the rotational diffusion properties of the head. This is determined from its fluorescence anisotropy decay curve [18]. The labelled S1 is excited with a short flash of polarized light and the polarization of the emitted fluorescent light is followed with time. If the S1 rotates within the life time of the excited state, then the emitted light will become depolarized (Fig. 4.6). From the

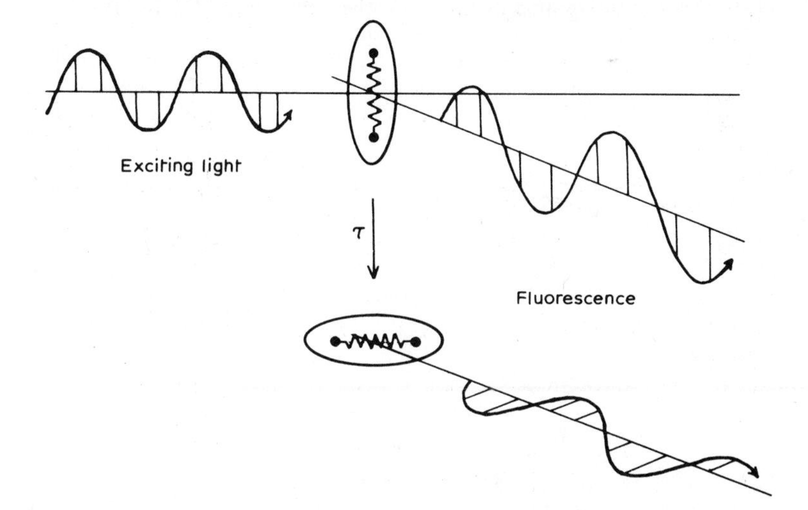

Fig. 4.6 The principle of fluorescence depolarization. Only molecules whose electronic dipole transition is parallel to the exciting light absorb the incoming light, but the polarization of the fluorescent light depends on the time of molecular rotation, τ, relative .o the lifetime of the excited state.

degree of depolarization, the rotational correlation time, τ, of S1 is calculated to be 100 ns, which is consistent with that of an ellipsoid of 5×15 nm. Notably, the τ values for HMM and myosin are only 2 to 3 times longer, indicating that each head is free to swing independently about its neck.

Another method for measuring rotational motion involves electron spin resonance (esr). A nitroxide spin label (Fig. 4.7a), containing a stable unpaired electron, is introduced by reaction with the SH1 thiol group. The label absorbs microwaves at specific frequencies dependent on the magnetic field. Three absorption peaks are observed because the nitrogen nucleus, about which the electron orbits, takes on three magnetic states (+ 1, 0, − 1), which complement the external field (Fig. 4.7b). The separation between the peaks, the hyperfine coupling (A), depends on the orientation of the nitroxide in the external field. In solution, the observed esr spectrum depends on the molecular rotation rate relative to the spread in A values (10^8 Hz). Subfragment 1 is static ($\tau = 100$ ns) on this time scale. Its broad esr spectrum represents the sum of signals for all orientations (Fig. 4.7c). Esr studies may be extended to give information about slower motions. The lifetime of the 'excited state' is about 10 μs, and at high microwave intensity the excess ground state population becomes depleted (saturation). The susceptibility to saturation at a specific frequency depends on whether the excited molecules can escape to a new orientation within their lifetime. Saturation transfer esr is sensitive to τ values in the range 0.1 ns to 1 ms. The technique has confirmed and extended fluorescence studies on myosin and shows that even in a myosin filament, the heads are free to rotate through a significant angle on the time scale of 1 μs [19].

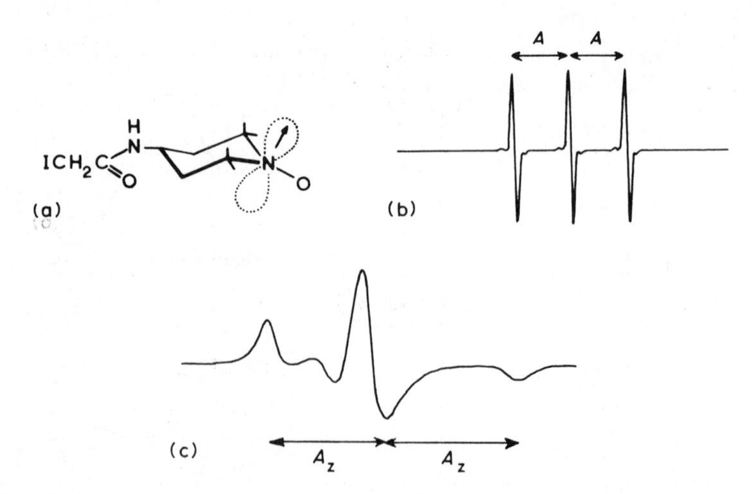

Fig. 4.7 (a) A cysteine-directed nitroxide spin-label, (b) the first derivative esr spectrum of a nitroxide group in an isotropic environment, (c) powder esr spectrum of an immobilized label reflecting a summation over all orientations.

These thermally-driven motions are likely to be important in muscle in allowing the myosin head to find an actin site at variable filament spacings. Moreover, it may be that the head remains rather rigid during contraction and filament sliding is achieved by a lever-like action at the actomyosin bond. The hinge regions would then act as universal joints in transmitting the motion to the backbone of the myosin filament.

The rate of ATP hydrolysis by myosin in the presence of physiological Mg^{2+} concentrations is rather slow, but then so is the ATP turnover by resting muscle. At 25°C the specific activity is about 25 nmol ATP per min per mg myosin, or 0.1 s^{-1} per active site. CaATP and KATP are hydrolysed at rates of 5 and 20 s^{-1} respectively, but these substrates are negligible in a muscle. Nevertheless, they provide a sensitive *in vitro* assay for myosin during its extraction and purification. Only the MgATPase is activated by actin and hence this reaction has received the most detailed attention.

4.4 Actin–myosin interactions

In the absence of MgATP, actin and myosin form a tight complex corresponding to the rigor state of muscle. Saturation transfer esr indicates that the myosin heads are rigidly attached and take on the comparatively slow flexing movements of the actin filament ($\tau = 1$ ms). Characteristic 'decorated filaments' are seen in the electron microscope when subfragment 1 is added to F-actin (Fig. 4.8). The arrowhead appearance of the attached subfragment 1 confirms that F-actin has a polarity. Indeed, this property is used to determine the direction of the actin filaments in muscle and non-muscle cells. The appearance of the decorated filament, along with studies of rigor muscle (Section 6.2), have

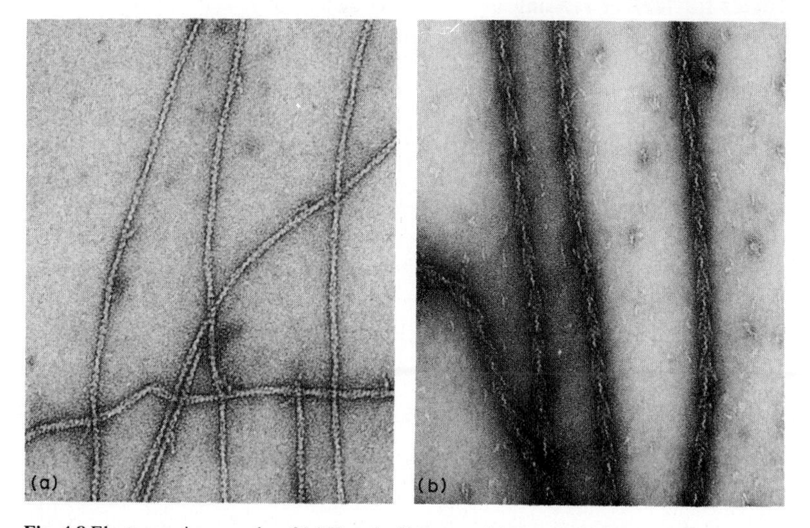

Fig. 4.8 Electron micrographs of (a) F-actin filaments, (b) F-actin filaments decorated with subfragment 1 heads. (Courtesy of Dr R. Craig.)

given rise to the notion of a 45° attachment angle of the heads. Although this concept is widely used in models of contraction, it should be remembered that the 45° appearance is due, in part, to viewing a 3-dimensional object in 2-dimensions [20].

In 0.5 M KCl actin and myosin form a viscous, turbid solution. On addition of MgATP, the solution clears indicating that the actomyosin complex dissociates. At low ionic strength actomyosin exists as a suspension. Again MgATP induces dissociation, but now the ATP is hydrolysed rapidly and as its concentration falls, the actin and myosin filaments reassociate and contract to form a precipitate (superprecipitation). While this may crudely represent contraction, the steric constraints involved make quantitative analysis difficult. Detailed kinetic studies have therefore been performed with subfragment 1 and heavy meromyosin, which remain soluble at low ionic strength.

It is instructive to compare the MgATPase rates observed *in vivo* (Section 2.5) with those measured in solution. The active site concentration in muscle is 240 μmol kg^{-1} (see Appendix), from which the ATPase activities of resting and isotonically contracting muscle are calculated to be $< 0.002\ s^{-1}$ and $6\ s^{-1}$ respectively. The MgATPase of frog skeletal subfragment 1 in solution at 0°C is $0.01\ s^{-1}$ and it is activated by actin to $4.5\ s^{-1}$ [21]. To a first approximation, the ATPase activities of the isolated proteins appear to reflect events in muscle. However, the resting ATPase rate *in vivo* is significantly lower, suggesting additional inhibitory factors may be present. Moreover, the comparison of the activated rates is not straightforward and the numerical agreement might not be as satisfactory as it first seems.

Topics for further reading

Harrington, W.F. (1979), Contractile proteins of muscle, in *The Proteins IV* (3rd edn) (Neurath, H. and Hill, R.L., eds), Academic Press, pp. 245–409. (A comprehensive review and reference source.)

5 Mechanism of ATP hydrolysis

5.1 Kinetic analysis

In searching for the molecular basis of contraction we need to relate chemical events with mechanical events. These must be quantitatively coupled in their extent (thermodynamics) and time course (kinetics). Unfortunately, chemical kinetic analysis of muscle itself presents practical and theoretical difficulties. In a muscle fibre the interaction sites are not freely diffusable and therefore do not combine by a simple bimolecular process. Also, once bound, the rates of subsequent steps depend on the tension borne by the filaments. Isolated proteins, on the other hand, form homogeneous solutions of defined concentration and their reactions may be initiated and terminated rapidly. Solution studies provide a basis for elucidating the catalytic mechanism of ATP hydrolysis and actin activation. Measurement under steady-state conditions reveal a correlation between the velocity of shortening of different muscle types and the ATPase rate of the actomyosin extracted from them (e.g. fast-twitch muscles have a high ATPase activity). However, our understanding of this correlation has been advanced through the application of transient kinetic techniques.

The simplest *in vitro* system relevant to muscle is the hydrolysis of ATP by subfragment 1. Catalysis is brought about after ATP binds to the active site and so Equation 5.1 represents a minimal pathway, where S1.X is an intermediate whose chemical nature is undefined.

$$\mathrm{S1} + \mathrm{ATP} \xrightarrow{k_1} \mathrm{S1.X} \xrightarrow{k_3} \mathrm{S1} + \mathrm{ADP} + \mathrm{Pi} \qquad (5.1)$$

If we assume the steps are irreversible then S1.X forms at the rate k_1 [ATP][S1] and decays at k_3 [S1.X]. Together with the conservation equation $[\mathrm{S1}]_0 = [\mathrm{S1}] + [\mathrm{S1.X}]$ these relationships allow calculation of the concentration of any species with time.

If the enzyme, S1, is present only in catalytic amounts then, soon after mixing, the formation of products becomes linear with time, i.e. a steady state is achieved where [S1.X] is constant. The rate of product formation depends on the proportion of S1 which is present as S1.X and hence

$V_{max} = k_3[\mathrm{S1}]_0$. Half saturation occurs when $k_1[\mathrm{ATP}] = k_3$ and therefore the Michaelis constant, $K_m = k_3/k_1$. The generality of the steady-state treatment is both its strength and weakness. Michaelis–Menten type kinetics are still observed if the reverse reactions are significant or if several intermediates are involved, but now the K_m and V_{max} may be defined by a combination of several rate constants.

Part of the ambiguity can be removed by focusing on the transient kinetics of the system, but this introduces technical problems of detection and time resolution. Concentration changes of the intermediates must be measured on the time scale of less than a single turnover. A comment about rate constants may be appropriate at this point. In the steady-state [S1 . X] is maintained and k_3 may be calculated from the maximum rate of Pi formation assuming that the S1 is pure and fully active. In the transient state, conditions may be arranged so that S1 . X is not regenerated and therefore it decays according to the equation [S1 . X] $\exp(-k_3 t)$, concomitantly releasing Pi. In this case, k_3 may be determined without knowledge of $[\mathrm{S1}]_0$. For such a first order reaction $k_3 = 0.69/t_{\frac{1}{2}}$ where $t_{\frac{1}{2}}$ is the time taken for [S1 . X] to fall by half any initial value. Another noteworthy characteristic is that the reaction is essentially (>99%) complete after 7 times $t_{\frac{1}{2}}$. Bimolecular binding reactions give pseudo first order kinetics if one reactant (e.g. ATP) is in large excess, because its concentration remains practically unchanged while the other reactant (e.g. S1) is depleted exponentially, with an observed rate constant $k_1[\mathrm{ATP}]$.

5.2 Myosin ATPase

There is considerable controversy as to whether the two heads of myosin are catalytically independent and functionally identical [22]. The pairing of the heads in the myosin molecule may be simply a reflection of the requirement of the heavy chains to form a stable coiled-coil. Single-headed myosin, prepared by controlled digestion, forms synthetic actomyosin threads capable of contraction, suggesting that pair-wise interaction of the heads is not fundamental to motility. The mechanism to be described refers specifically to subfragment 1 from rabbit skeletal muscle at 20°C.

The slow turnover of MgATP by subfragment 1 serves as a model for relaxed muscle and also provides the basis for an understanding of actin activation. ATP hydrolysis can be followed discontinuously by denaturing the myosin with acid and assaying the ADP or Pi concentration. Alternatively, ADP or Pi can be monitored continuously by using a linked-assay system. There is no chromophoric change associated with the ATP hydrolysis, but certain enzymes utilize the products in conjunction with the NAD–NADH redox reaction. Provided these coupling enzymes are in sufficient concentration the rate of NADH conversion, conveniently monitored by the change in absorbance at 340 nm, will be limited by the myosin-catalysed production of free ADP or Pi. For example:

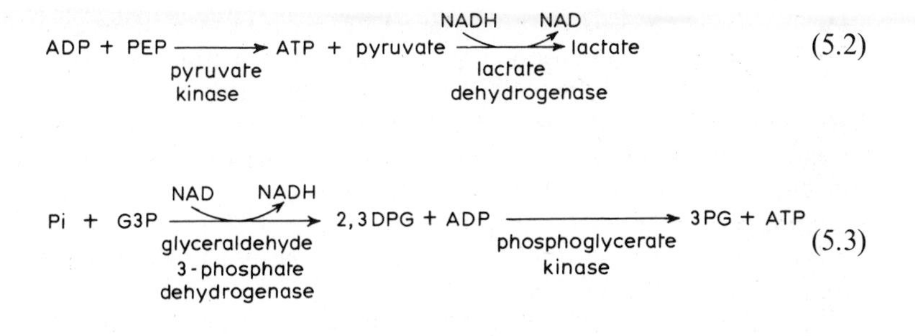

where G3P is glyceraldehyde 3-phosphate, 2,3DPG is 2.3-diphosphoglycerate and 3PG is 3-phosphoglycerate. In both reactions ATP is regenerated, thereby maintaining the initial ATP concentration.

During the steady-state phase of the reaction, the linked assay and acid quench methods yield the same rate ($0.1\ s^{-1}$) as expected. However, extrapolation of the data back to zero time reveals that, in the case of the acid quenching method, there is a rapid production of products, almost equivalent to the concentration of the S1. To analyse this aspect further it is necessary to mix the reactants rapidly and to follow the time course on the millisecond time scale. This is achieved by using a quenched-flow or stopped-flow apparatus (Fig. 5.1).

In the simplest form of the quenched-flow apparatus the two reactants are expelled through a mixing jet and driven along a calibrated tube into the acid quenching solution. The age of the reaction at the time of quenching is varied by altering the speed of the syringe drive or the length of the tube. The concentration of the products at each time point

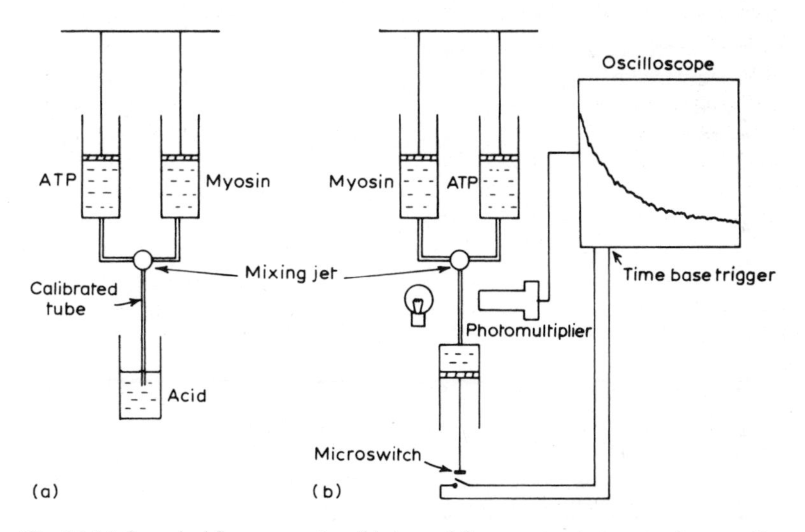

Fig. 5.1 (a) Quenched flow apparatus, (b) stopped-flow spectrophotometer for recording rapid absorbance or turbidity changes.

allows a progress curve to be constructed, i.e. it is a discontinuous assay. In the stopped-flow apparatus the reactants, after mixing, pass into a third syringe whose plunger meets a stop incorporating a microswitch. The switch triggers the time base of an oscilloscope, which then continuously records the aging of the arrested mixture spectrophotometrically. Both techniques are limited by the time (about 1 ms) taken to mix and transfer the reactants to the quenching or observation chamber. Reactions with rate constants $> 1000\ s^{-1}$ will be practically complete within the dead-time of the instrument.

Fig. 5.2 shows the observed time course of Pi production when subfragment 1 is mixed with an excess of ATP in these rapid reaction devices. Note that the linked-assay system detects only free Pi, whereas the acid quench measures, in addition, any labile bound Pi. The data of Fig. 5.2 argue that the rate-limiting step occurs after ATP hydrolysis at the active site. Similar results are obtained if ADP production is monitored. Equation 5.1 may therefore be extended to Equation 5.4 as the minimum description of the pathway, where $k_3 = 0.1\ s^{-1}$.

$$\mathrm{S1 + ATP} \xrightarrow[\substack{\text{variable}\\ \text{(fast)}}]{k_1} \mathrm{S1.ATP} \xrightarrow[\text{fast}]{k_2} \mathrm{S1.ADP.Pi} \xrightarrow[\text{slow}]{k_3} \mathrm{S1 + ADP + Pi} \qquad (5.4)$$

In the quenched flow experiment (Fig. 5.2) the transient phase is attributed to the formation of S1.ADP.Pi which decomposes in strong acid. The observed first order rate constant for this process k_{obs} is either k_1[ATP] or k_2 whichever is the slower. At low [ATP], k_{obs} is linearly dependent on [ATP] and indicates that $k_1 = 10^6 M^{-1}s^{-1}$. At high [ATP], k_{obs} is independent of [ATP] and shows that $k_2 = 100\ s^{-1}$. Equation 5.4 has been expanded further to take into account the time course of myosin fluorescence and pK changes which are sensitive to ATP binding and the release of ADP and Pi [22]. Equation 5.4, however, remains an adequate description for our purpose, although we need to explore the reversibility of the pathway. So far, we have assumed k_{-1} is negligibly small, without any justification.

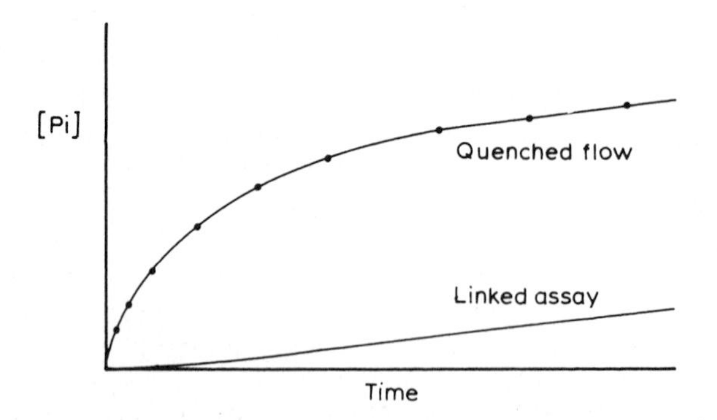

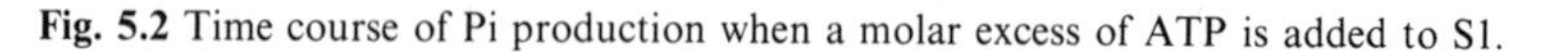
Fig. 5.2 Time course of Pi production when a molar excess of ATP is added to S1.

An enzyme cannot alter the equilibrium of a reaction. That does not mean to say individual steps may not be perturbed, but that the overall balance $K_1 . K_2 . K_3 = K_{eq} = 10^6$ M must be maintained. Supposing a single turnover of ATP hydrolysis is examined by mixing an excess of S1 (say 10 μM) with γ-^{32}P labelled ATP. The binding phase will be pseudo first order with $k_{obs} = k_1[S1] = 10$ s^{-1}. Once formed S1.ATP will immediately hydrolyse to S1.ADP.Pi ($k_2 = 100$ s^{-1}). Hence after 0.5 s the binding and hydrolysis steps will be complete but S1.ADP.Pi decay will have barely started. In practice when the reaction is quenched at this time, 90% of the ^{32}P label is recovered as Pi and 10% remains in the ATP. This may be explained if the hydrolysis step is reversible with $K_2 = k_2/k_{-2} = 9$. Addition of an excess of unlabelled ATP at 0.5 s has no effect on the subsequent hydrolysis of the labelled S1.ATP complex showing that $k_{-1} < k_3$. The reversibility of the hydrolysis step has been confirmed by isotope exchange measurements [23]. Studies using $H_2^{18}O$ indicate that water is incorporated reversibly at step 2, so that the Pi which is eventually released from S1.ADP.Pi derives all its oxygen atoms from the solvent. Moreover, addition of high concentrations of S1 to ADP and Pi results in a small but significant production of S1.ATP, and indicates that the irreversibility of the overall ATPase reaction is largely attributed to ATP binding. The ATPase reaction results in the inversion of the stereochemical configuration of the oxygen atoms of the terminal phosphate group and suggests that step 2 is a simple in-line attack of water [24]:

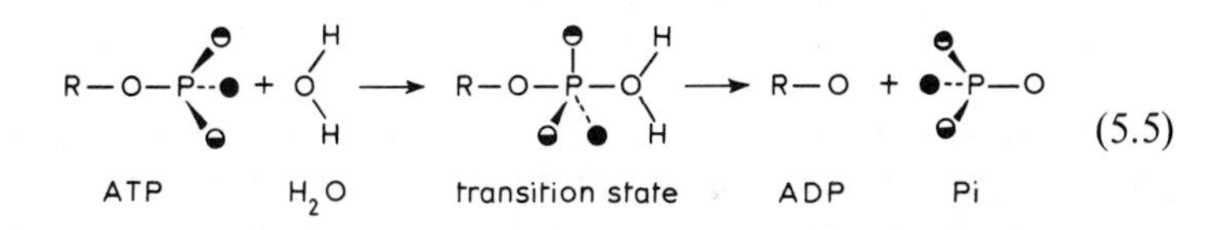

If a phosphorylated myosin intermediate was involved, the resultant double-displacement reaction would lead to an overall retention of configuration by the product Pi.

5.3 Actin activation

Actin activates the myosin MgATPase only when present in the filamentous F-form. However, conventional solution kinetics still apply in the case of its interaction with isolated S1 heads. Steric complications are introduced when the heads are tethered as in HMM and these are further compounded in the myosin filament. In practice the interaction of HMM is rather similar to S1, a finding which is consistent with the high degree of flexibility about the S1–S2 link. Myosin, on the other hand, shows a lower degree of activation presumably because the probability of the filaments coming together in the correct orientation is rather low and, even if they do, the head periodicity will not match that of the actin sites.

Another problem encountered in kinetic analysis concerns the high protein concentration of muscle (about 100 mg ml^{-1} actomyosin) which cannot be achieved in solution. A compensatory, but not completely satisfactory course of action is to measure the ATPase at very low ionic strength (0.01 M) where the affinity of actin for the subfragment 1 intermediates is enhanced. Under these conditions the V_{max} of the actin-activated S1 ATPase is about 20 s^{-1} and the K_m for actin is 40 μM [22]. Note that actin concentrations are expressed in terms of the number of G-monomers, although significant activation is achieved only with the F-form.

Steady-state kinetics indicate that actin behaves as a simple activator, but the process seems paradoxical in that on adding ATP to acto-subfragment 1, the turbidity and viscosity drop practically to the values expected for the dissociated proteins. In other words, actin achieves activation by a transient interaction with subfragment 1, but this is sufficient to bypass the slow products release step controlled by k_3. Equation 5.6 represents the potential interactions of actin with the S1 intermediates of Equation 5.4.

$$\begin{array}{ccccccc}
\text{A.S1 + ATP} & \overset{k_4}{\rightleftharpoons} & \text{A.S1.ATP} & \overset{k_5}{\rightleftharpoons} & \text{A.S1.ADP.Pi} & \overset{k_6}{\rightleftharpoons} & \text{A.Si + ADP + Pi} \\
k_{-a}\uparrow\downarrow k_a & & k_{-b}\uparrow\downarrow k_b & & k_{-c}\uparrow\downarrow k_c & & k_{-a}\uparrow\downarrow k_a \\
\text{S1 + ATP} & \overset{k_1}{\rightleftharpoons} & \text{S1.ATP} & \overset{k_2}{\rightleftharpoons} & \text{S1.ADP.Pi} & \overset{k_3}{\rightleftharpoons} & \text{S1 + ADP + Pi}
\end{array} \qquad (5.6)$$

In the absence of ATP, subfragment 1 binds tightly to actin (k_a = 0.1 μM) with a stoichiometry of one head per actin monomer. Thus a mixture of 10 μM S1 and 10 μM actin will exist essentially as A.S1 (termed a rigor complex). With this preparation in one syringe of a quenched-flow apparatus (Fig. 5.1), and ATP in the other, Pi production may be followed as before (cf. Fig. 5.2). The time course of the transient phase is very similar to that for S1 alone, but the steady-state is now increased to about 5 s^{-1} (note [A] < K_m^{actin}. Increasing the [ATP] increases the observed rate constant of the transient phase which reaches a maximum of about 100 s^{-1} as with S1 alone. The significance of this finding is clarified when turbidity changes are followed in a stopped-flow apparatus (Fig. 5.1). At low ATP concentrations the turbidity of the acto-subfragment 1 drops with the same time course as the transient Pi production, but the rate constant increases linearly with [ATP] until it becomes too fast to measure (> 1000 s^{-1}). In terms of Equation 5.6 it appears that k_4 has a similar value to k_1 (about 10^6 M^{-1} s^{-1}) but once formed A.S1.ATP dissociates to S1.ATP (k_b > 1000 s^{-1}) before hydrolysis takes place, so that Pi formation is limited by k_2 regardless of the presence of actin. Actin then reassociates (k_{-c}) to displace the products rapidly and achieve an overall activation ($k_6 >> k_3$). Note that the tight binding of ATP to S1 is used, in effect, to drive the dissociation of the A.S1 complex. This pathway predominates at low actin concentrations as shown by Lymn and Taylor [25]. Assigning the rate-limiting step in the actin-activated pathway is difficult because it requires high actin concentrations. Actin rebinding is second order (i.e. k_{-c} $[A] \rightarrow \infty$) so that product release from A.S1.ADP.Pi might be expected to become rate limiting (i.e. $k_6 = 20\ s^{-1}$) at saturating actin. However, as V_{max} is approached by increasing the actin concentration, the turbidity indicates that a significant proportion of the S1 remains dissociated from actin. This observation led to the proposal of a refractory state form of S1.ADP.Pi which undergoes a rate-limiting transition before actin rebinds. Further studies showed that at high actin concentration A.S1.ATP dissociation to S1.ATP, although rapid, is incomplete and that direct hydrolysis to A.S1.ADP.Pi may occur without dissociation [26]. Despite these complications the scheme of Equation 5.6 allows the following qualitative argument to be formulated:

1. The nature of the nucleotide at the myosin active site changes during the ATP hydrolysis cycle.
2. The affinity of actin for myosin depends on the state of the bound nucleotide ($K_a < K_c < K_b$, note that these are dissociation equilibrium constants).
3. Hence, there will be a tendency for S1 heads to bind to actin more readily in one state and detach more readily in another.
4. Assuming that the attached states differ in some structural characteristic (e.g. attachment angle) the heads will preferentially sweep out an arc in one direction (actin is polar) before detaching during each cycle of ATP hydrolysis. Tethering the heads to a filament backbone will therefore cause motion of the myosin filament relative to the actin filament.

This is the basic framework envisaged by Lymn and Taylor [25]. In discussion of this and subsequent models S1 has been equated with the myosin head, M, and ATP and ADP.Pi are abbreviated as T and D respectively. Lymn and Taylor specifically identified detachment with A.M.T→M.T and attachment with M.D→A.M.D. Angle changes are assumed to occur between the A.M.D, A.M and A.M.T states (Fig. 5.3). By analogy with the rigor state (Section 4.4) the A.M intermediate is depicted with an attachment angle of 45°.

As noted above, the Lymn–Taylor mechanism is not an obligatory pathway in solution, let alone muscle. M.T faced with an adjacent actin may bind, while A.M.D may dissociate, particularly if the filament is under tension. The extent to which these solution studies are of value in formulating crossbridge action is considered in Section 6.4. Nevertheless the phenomenon of actin activation of ATP hydrolysis at the enzymic site of myosin, and the identification of subfragment 1 as part of the crossbridge, provides irrefutable evidence that the latter is, at minimum, a key accomplice in the contractile event.

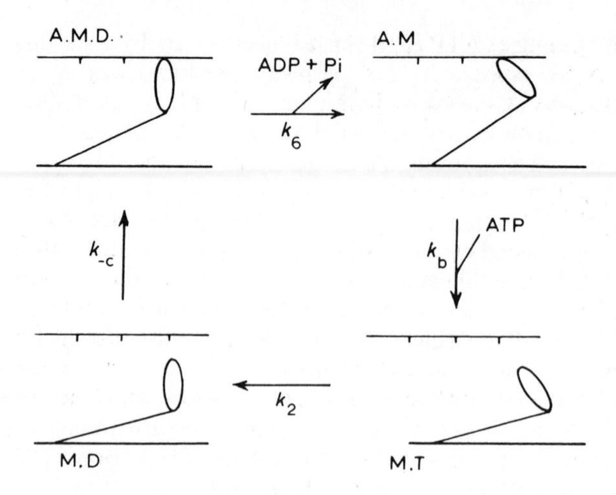

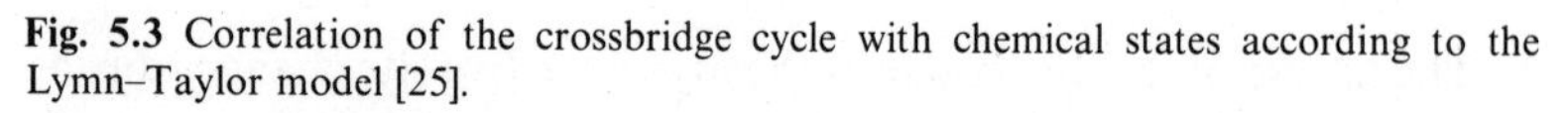

Fig. 5.3 Correlation of the crossbridge cycle with chemical states according to the Lymn–Taylor model [25].

Topics for further reading

Fersht, A. (1977), *Enzyme structure and mechanism*, Freeman and Co., San Francisco. (An introduction to kinetic methods.)

Trentham, D.R., Eccleston, J.F. and Bagshaw, C.R. (1976), *Q. Rev. Biophysics*, **9**, 217–281. (A review of ATPase mechanisms.)

6 Molecular basis of contraction

6.1 What makes the filaments slide?

Although the myosin heads are undoubtedly involved in the mechanism of filament sliding, they are not necessarily the mechanical agents for force generation. We may consider three general hypotheses by which relative motion of the filaments could be achieved: 1. maximum interaction, 2. lateral expansion, 3. independent force generators [27].

The first hypothesis is illustrated by the example in Fig. 6.1 where the filaments shorten to achieve maximum neutralization of filament charge (i.e. they behave as capacitor plates). Note that the actual driving force occurs only at the tips of the filaments which bring new + and − charges together. The tension generated is therefore independent of overlap and drops to zero at complete overlap. This is contrary to experimental findings.

The lateral expansion hypothesis may be illustrated by a model in which the thick and thin filaments develop like charges. The resultant lateral repulsion in conjunction with the constant volume of the muscle fibre would cause longitudinal shortening. However, skinned fibres do not adhere to the constant volume relationship yet they generate practically the same tension as intact fibres, so indicating that lateral expansion is not a causal phenomenon.

The independent force generator theory, as its name implies, considers sarcomere contraction to result from an array of elements, each capable of producing a force in the direction of shortening. The idea was born before the crossbridge was identified as a physical entity in electron micrographs, and indeed draws on physiological evidence from the 1920s. In 1957 A.F. Huxley [28] published an influential model in this category which accounted for the steady-state force, velocity and energy expenditure of a muscle (Figs 2.4 and 2.7). A 'side-piece' elastically connected to one filament was proposed to attach to the other filament in a strained state with a moderate rate constant, but to detach rapidly, under the influence of ATP, only when in an unstrained or compressed state. The strained state was assumed to arise from thermal motion but ATP hydrolysis is essential to provide a net vectorial reaction. The

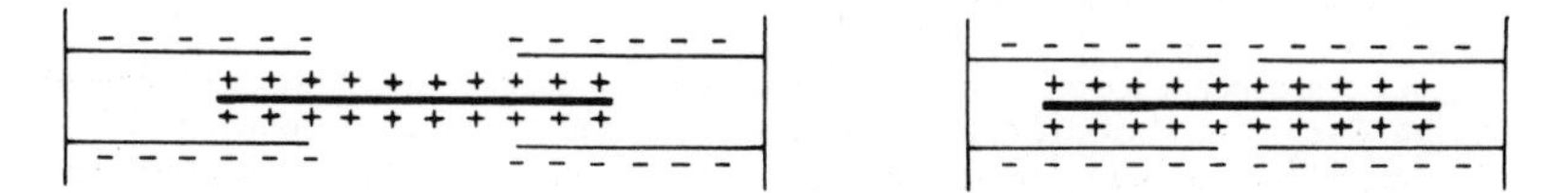

Fig. 6.1 A theory of muscle contraction based on maximum interaction.

coupling between muscle shortening and ATP hydrolysis (Fig. 2.7) arises from the variable rate constant for detachment. When the filaments are fixed, isometric tension is generated by the attached strained side-pieces which accumulate because of their unfavourable dissociation rate. Consequently, the cycling rate and ATPase rate is less than in isotonic contraction where the side pieces can move into an unstrained position. The decrease in ATP utilization at high velocities of contraction is a consequence of the attachment step. Many side-pieces do not have time to complete their interaction with the actin filament before being actively pulled off. In effect contraction is driven by a few side pieces whose thermal energy happens to lie at the upper end of a Boltzmann distribution. This does not defy thermodynamics because such high velocities are only achieved at low loads, where the actual work output approaches zero (Fig. 2.4). The model is a kinetic one, but soon after its proposal crossbridges were seen in electron micrographs emanating from the thick filament and these were identified as the sites of ATP hydrolysis. The crossbridge was therefore taken to be the side-piece responsible for cyclic attachment. However, the identity of the elastic element remains in question.

If crossbridges do act as independent force generators then the net isometric tension should depend on the number attached and hence it should be proportional to the degree of overlap between the thick and thin filaments (cf. maximum interaction hypothesis). While the expected correlation between tension and length of an intact muscle is observed to a first approximation, a quantitative comparison is complicated by passive elasticity and non-uniformity in sarcomere lengths. These problems were reduced by examining the central portion of a single muscle fibre whose length was controlled by a feedback circuit (Section 6.4). When corrected for the passive elasticity present in the unstimulated fibre, the initial isometric tension developed is in good agreement with that expected from the degree of overlap (Fig. 6.2). Note

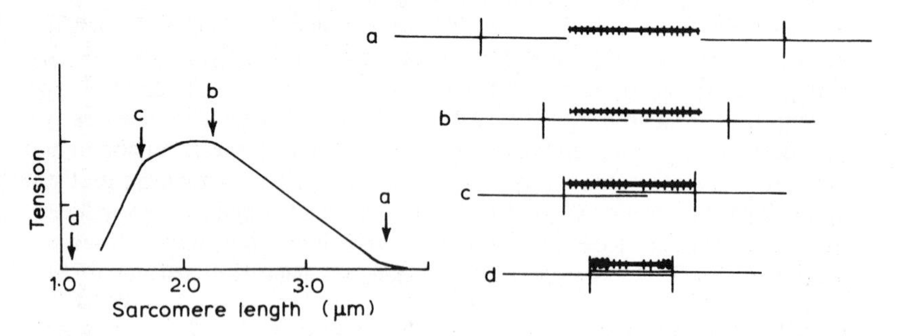

Fig. 6.2 The dependence of active tension generation by a frog muscle fibre on sarcomere length. The passive elasticity of a relaxed muscle rises steeply above a sarcomere length of 2.5 μm and this has been subtracted from the total tension of the stimulated muscle. The positions indicated correspond to sarcomere lengths of (a) 3.65 μm, (b) 2.2 μm, (c) 1.65 μm, (d) 1.05 μm. (From Gordon *et al.* [29].)

that an isolated muscle can contract beyond the point of maximum overlap (supercontraction), although tension is reduced, presumably for steric reasons. The ends of the actin filaments first overlap to form a visible Cm contraction band and then the thick filaments buckle at the Z-line to give a Cz contraction band.

The negative length–tension relationship at partial filament overlap gives rise to an instability which would be disastrous if parallel passive elasticity was totally lacking. If one sarcomere shortened slightly at the expense of its neighbour then the process would escalate. However, the membrane, with its surrounding connective tissue, acts like the mesh in a stocking and resists overstretching. Also there is some evidence that the thick filaments are connected to the Z-line by an elastic protein, termed connectin. Nevertheless, isolated fibres are prone to creep owing to this instability and in demembranated preparations (e.g. glycerinated fibres) only a limited number of contraction–relaxation cycles can be achieved before clots of supercontracted material appear.

A second finding in support of independent force generators is that the maximum velocity of contraction is independent of the degree of overlap and hence, of the number of potential interaction sites available in each sarcomere. It follows from this relationship that a muscle with a large number of short sarcomeres can contract faster, but develop less tension than one with fewer, longer sarcomeres. While this correlation is found in some arthropod muscles, other factors, such as the myosin isotype, may be involved. Although the independent force generator hypothesis provides a viable mechanism for filament sliding, it reopens the question as to the nature of the force on a smaller scale.

The properties of asynchronous insect flight muscle appear very different from skeletal muscle, although the mechanism of force generation at the molecular level is probably rather similar. Its I-bands are very short and the thick filaments are connected to the Z-line by an elastic protein which gives a high resting stiffness. When the stimulated muscle is stretched it responds with a delayed increase in tension. This allows insect flight muscle to drive an inertial load (e.g. a wing and ligament) at its natural frequency of oscillation, in the same way as we maintain a swing in motion. The muscle only shortens by a few percent of its rest length which is the same order of filament sliding that might be achieved by a crossbridge cycle.

6.2 Electron microscope studies

The problem of muscle contraction has become one of defining the structure and movements of the crossbridge. In relaxed insect flight muscle crossbridges appear to project outward ($\sim 90°$) at a periodicity of 14.5 nm reflecting their packing in the thick filament (cf. Fig. 3.4). In rigor, the crossbridges attach to the thin filament at an angle of about 45° and appear at intervals of 38.5 nm, which reflects the position of suitable actin sites (Fig. 3.3). These observations led to the idea that the crossbridges might attach at some angle around 90°, then move to 45°, so causing the filaments to slide by a few nanometres. Can we synthesize a 'movie' of the crossbridge cycle from a collection of electron micrograph 'stills'? There are a number of formidable problems. During contraction itself crossbridges move asynchronously and only the most stable states (e.g. rigor) can be trapped readily. In addition the crossbridges are liable to distortion during the fixing, sectioning and staining procedures. However, much of the disorder introduced is random and the underlying regularities can be retrieved by optical filtration. The principle of the technique is shown in Fig. 6.3.

The electron micrograph (o) is illuminated with a parallel monochromatic light beam and an image (i) formed with the lens (l). Each point on the micrograph acts as a secondary light source from which a new wavefront emerges. The wavefronts complement each other in the forward direction and in a conventional ray diagram we depict this as a parallel beam which, after refraction by the lens, passes through its focal point. Periodic features of the micrograph cause the wavefronts to come into step at certain other angles, represented by the diffracted ray (---) in Fig. 6.3. Diffracted rays converge at the focal plane (f), but off-axis, to give a pattern of bright spots. Light scattered by aperiodic features results in a diffuse disc at the focal plane. To filter the image a mask is made by piercing holes in a foil sheet corresponding to the diffraction pattern. With the mask in place the background scatter from the aperiodic features is eliminated and the regular features of the micro-

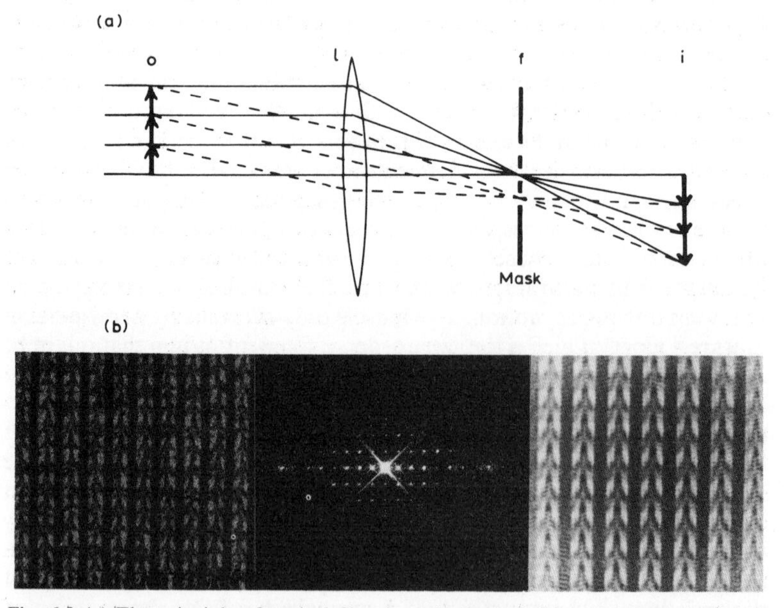

Fig. 6.3 (a) The principle of optical filtration. (b) Optical diffraction pattern and the corresponding filtered image of an electron micrograph of insect flight muscle in rigor. (Courtesy of Drs M. and M.K. Reedy.)

graph are enhanced. An unambiguous image cannot be retrieved from a photograph of a diffraction pattern because phase information is lost.

Optical filtration of electron micrographs of insect flight muscle in rigor clearly reveals the 45° attachment angle of the crossbridges. Vertebrate skeletal muscle is less well preserved in electron micrographs but the 45° attachment angle is apparent in decorated thin filaments (Fig. 4.8). Although electron micrographs are only 2-dimensional the helical nature of actin allows us to view the subfragment 1 at different angles and to reconstruct a 3-dimensional image. In such an analysis subfragment 1 appears the shape of a cupped hand. The distal end, corresponding to the fingers, projects outward on successive actin subunits and gives rise to the characteristic arrowhead pattern [20]. States other than rigor are difficult to examine and these will be considered subsequently in conjunction with other techniques.

6.3 X-ray diffraction studies

The problems of electron microscopy arise from the need to examine a stained specimen *in vacuo*. In principle an X-ray microscope would be better because it would reveal atomic detail continuously and non-destructively. Alas, because of the difficulty of refocusing X-rays after diffraction by the object, the technique has not yet been developed to its full potential. However, as we have seen (Fig. 6.3), a diffraction pattern contains all the information required to define an image. For periodic structures which give rise to discrete spots, an image can be reconstructed mathematically provided the position, intensity and phase of each is known. Phase determination is a problem. The breakthrough which allowed the high resolution structure of a protein crystal to be solved was the introduction of a heavy atom as a reference point. For a simple pattern without phase information, we can guess at a reasonable structure, compare its computed diffraction pattern and then refine the guess.

Muscle is not as ordered as a crystal, but nevertheless it gives a characteristic diffraction pattern which enables periodic structures to be investigated [30]. At high angles of diffraction (corresponding to <2 nm spacing) the predominant pattern arises from the α-helical rod of myosin. Prior to the sliding filament theory, contraction was thought to involve a change in filament length and a helix–coil transition in the rod appeared a possible mechanism. However, the high-angle diffraction pattern remains unchanged during contraction. At lower angles of diffraction, information is revealed about the periodicities of the filaments and their projections. Some of these spots do change on contraction. Considerable effort has therefore been directed towards identifying the structures responsible for the pattern, by correlation with periodic features observed in electron micrographs and investigation of paracrystals (ordered filaments) of purified muscle proteins. Fig. 6.4 shows the arrangement for recording a low angle diffraction pattern of a muscle.

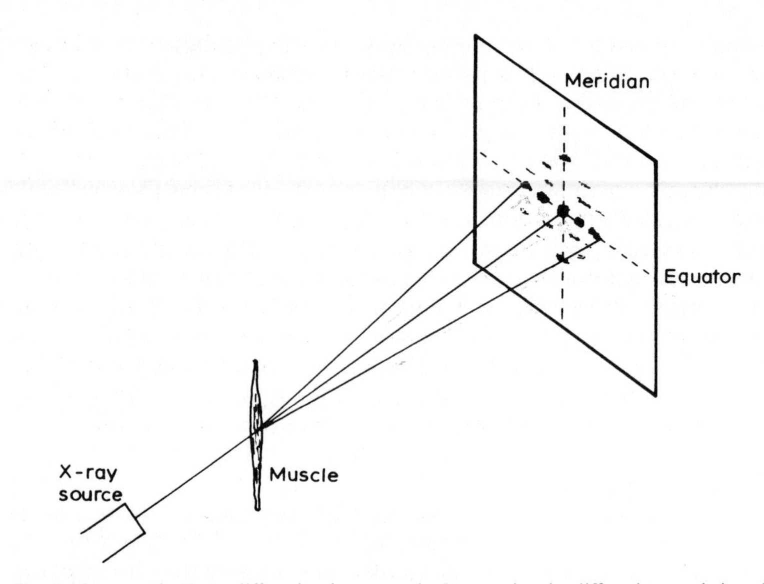

Fig. 6.4 Low angle X-ray diffraction by a muscle. In practice the diffraction angle is only about 2°, but this is exaggerated in the diagram.

The two main spots on the equator arise from the 1.0 and 1.1 diffraction planes of the hexagonally arranged thick and thin filaments (Fig. 6.5). Their position therefore monitors the lateral expansion of muscle on shortening. More revealing are their intensity changes. On going from the relaxed to the rigor state the 1.1 spot increases and the 1.0 spot decreases showing that electron density (presumably crossbridges) moves from the thick to the thin filament.

Contracting muscle is more difficult to examine because it can develop

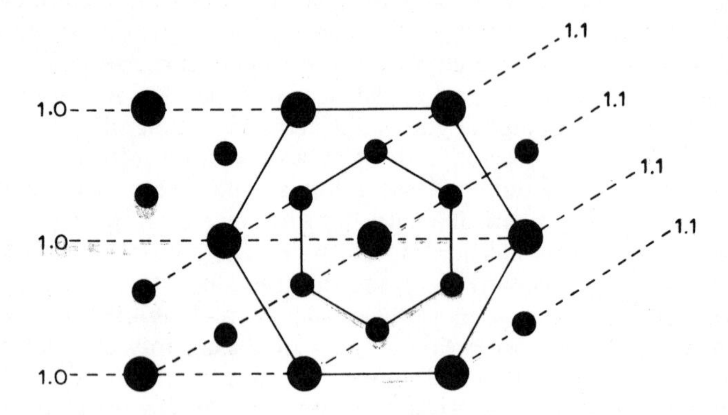

Fig. 6.5 Cross-section of vertebrate skeletal muscle showing the double hexagonal arrangement of the thick and thin filaments. The equatorial X-ray diffraction pattern arises from 1.0 and the 1.1 planes. Different muscle types show great variation in the ratio and arrangement of thick and thin filaments.

isometric tension for only a few seconds before fatiguing. Although diffraction is effectively an instantaneous process the proportion of X-ray photons scattered is small so that the time resolution is controlled by the need to accumulate a detectable signal. The first patterns of contracting muscle were obtained by synchronizing the exposure of a photographic plate with excitation and a discernable pattern was built up by repeated stimulation. The minimum exposure times have shown a steady decrease with increasingly powerful X-ray sources and more sensitive detection systems. Using intense synchrotron radiation and an electronic position-sensitive X-ray counter the stronger peaks can now be detected within 100 ms [31]. The signal from the counter may be stored in successive time-bins of a computer memory, allowing a time course to be built up by repeated stimulation. A frog sartorius muscle will survive several hundred contractions which enables the X-ray pattern to be obtained with about 1 ms time resolution.

$$\text{time resolution} = \frac{\text{minimum exposure time}}{\text{number of repeated contractions}} \tag{6.1}$$

The changes in the 1.0 and 1.1 reflections indicate that crossbridges move out towards the thin filament a little ahead of tension generation and their final ratio approaches that of the rigor state. The fraction of crossbridges attached in isometric contraction, based on these intensities, is between 40–80%. The calculation is inexact because disposition of electron density depends also on the shape and angles of attachment of the crossbridges. Although these intensity changes are good evidence that crossbridge movement is a prerequisite for contraction, the equatorial spots are sensitive only to lateral movements. In the crossbridge model implied in Sections 5.3 and 6.1, it was assumed that crossbridges move longitudinally; for supporting evidence we need to look at the meridional spots and layer line diffraction pattern. To understand these reflections we may consider the form of the diffraction pattern given by electron density centred at points on a helix (Fig. 6.6).

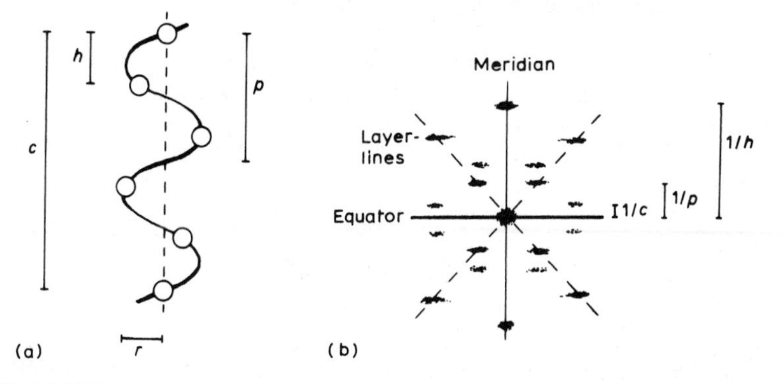

Fig.6.6 (a) A non-integral helix in which $c = 2p = 5h$, i.e. there are 2.5 subunits per turn. (b) The corresponding diffraction pattern.

The distance of the first layer line from the equator provides a measure of the true repeat, c, while the distance of the first layer line to appear on a 'cross' passing through the origin gives the pitch, p. For an integral helix $c = p$. Information about the radial distribution of mass, r, is contained in the distance of the layer line intensity from the meridian. The first spot to appear on the meridian provides the subunit repeat distance, h. The interpretation of the diffraction pattern of muscle itself is more complicated because the components are distributed in several multiple helical arrays [30].

Relaxed skeletal muscle shows a meridional spot at 14.3 nm and layer lines at 42.9 nm which arise from the helical arrangement of crossbridges about the thick filament (Fig. 6.7).

The pattern cannot be interpreted unambiguously but it indicates that the crossbridges are arranged on n helical strands with a pitch of $n \times 42.9$ nm and an axial repeat of 14.3 nm (cf. the 14.5 nm repeat observed in scallop filaments in Fig. 3.4). The weight of the evidence favours $n = 3$ for vertebrate skeletal muscle (Fig. 6.8). In particular its thick filaments are prone to fray into 3 subfilaments indicating a 3-fold symmetry in the packing of the backbone [32]. Invertebrate muscle thick filaments have a larger diameter and the crossbridges may be disposed on a 4 or more stranded helix. However, they retain a similar axial spacing (14.5 nm). Note that the interpretation of the X-ray pattern of skeletal muscle does not imply that the myosin tails are actually twisted together in a triple helix but merely describes the favoured position of the crossbridge in time and space. While the 14.3 nm spot is the strongest meridional reflection, the presence of other 'forbidden' meridional spots (e.g. 21.5 nm) show that the crossbridge spacing is not perfectly regular (Fig. 6.7). Nevertheless the X-ray data indicate that in relaxed muscle the position of the crossbridges is determined essentially by their attachment to the thick filament backbone. This contrasts markedly with the situation in rigor where the crossbridges take on the spacing of the thin filament.

With the help of actin paracrystals and electron micrographs, the helical nature of the thin filaments has been established in detail. The diffraction pattern, with a meridional spot at 2.7 nm and layer lines at 5.1 and 5.9 nm, arises from the double strand of G-monomers with a diameter of 5.5 nm, in which the chains are staggered by half a subunit (i.e. 2.7 nm) with respect to each other (Fig. 4.2). The two chains are twisted together with a long pitch of 74 nm and therefore crossover at 37 nm to give a corresponding layer line. In such a model we can trace out tightly coiled right- and left-handed helices (known as genetic or primitive), which take in the alternate staggered subunits of each chain and it is their pitches that give the 5.1 and 5.9 nm layer lines.

When a skeletal muscle goes from the relaxed to the rigor state the 5.1, 5.9 and 37 nm layer lines are all intensified, while the 42.9 nm layer line disappears. These results suggest that the crossbridges bind and take on the helical characteristics of actin at the expense of losing their thick filament helical symmetry. However, the 14.3 nm meridional is reduced

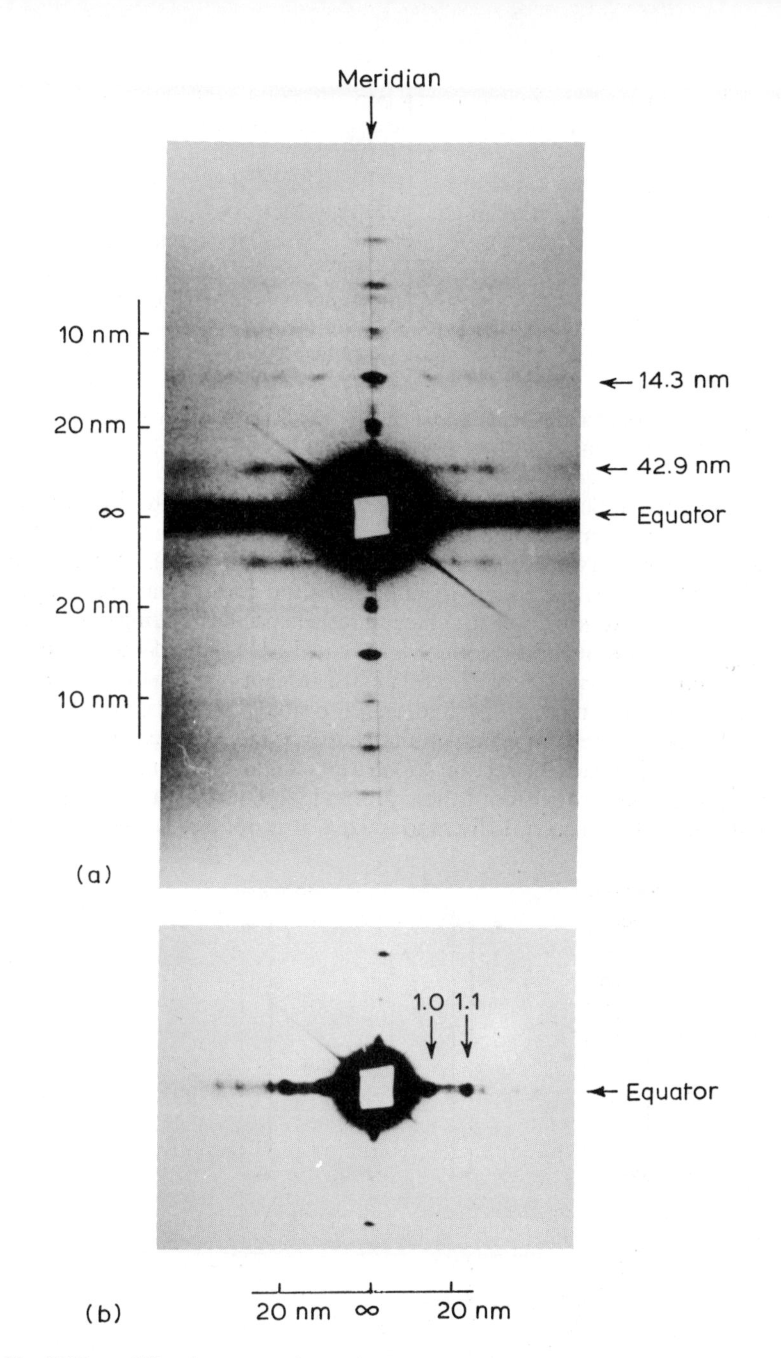

Fig. 6.7 X-ray diffraction pattern from relaxed frog muscle. In (a) the photographic plate was exposed to reveal the layer lies, while in (b) a shorter exposure was used to resolve the equatorial spots. (Courtesy of Dr H.E. Huxley.) Note that a spot is named in terms of the real space of the diffracting object, but it appears at a reciprocal position in the pattern.

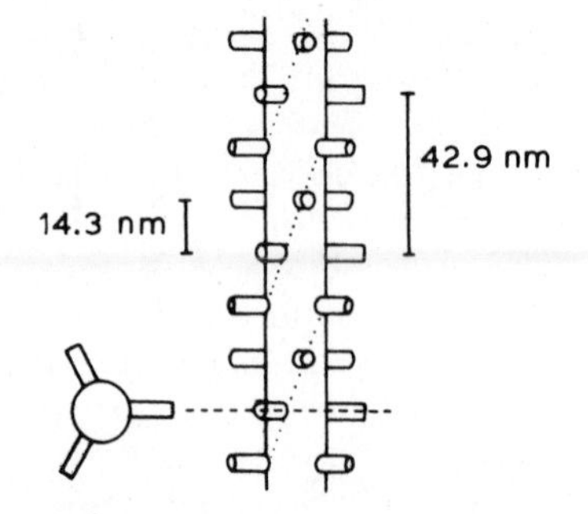

Fig. 6.8 A model of a vertebrate skeletal muscle thick filament, according to Squire, in which the crossbridges lie on a 3-stranded helix. (From Offer [33].)

but not lost, indicating that crossbridges can find actin sites by a radial (lateral) movement accompanied by a slewing about the azimuth, rather than extensive axial (longitudinal) movement (Fig. 6.9). Despite the mismatch in the actin and myosin periodicities and the variable lateral spacing, the heads are able to achieve a rather specific bonding with actin because of the flexibility in the S1–S2 and LMM–HMM link regions.

Spectroscopic studies of skeletal muscle in rigor confirm that there is a rather limited angular distribution of heads. These studies involve attaching a fluorescent group or nitroxide spin label to the SH1 cysteine of the myosin head (see Section 4.3) in a glycerinated fibre and exploiting the directional properties of these probes [34]. On relaxation the probe dipoles become directionally randomized. It is not clear to what extent

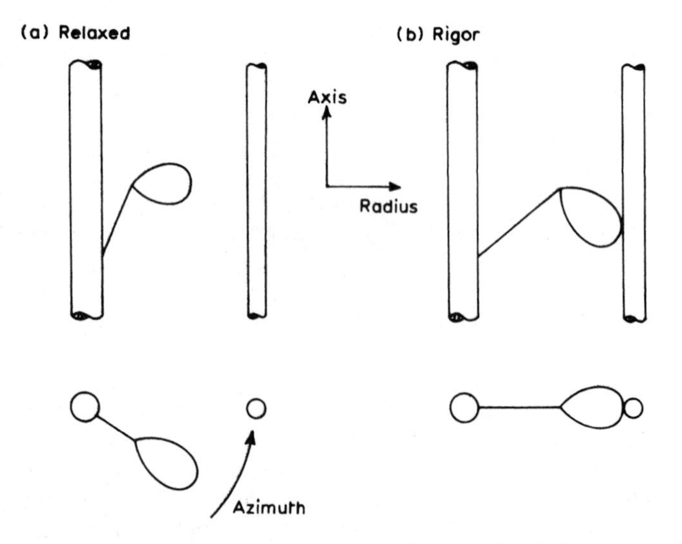

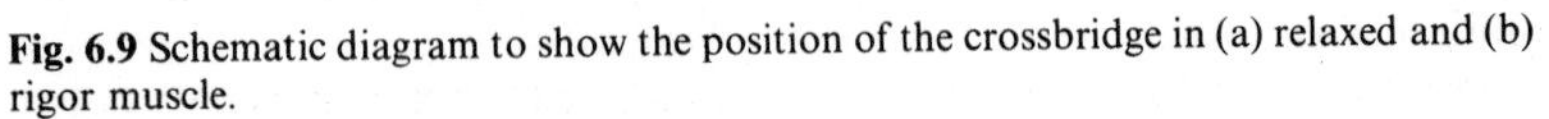

Fig. 6.9 Schematic diagram to show the position of the crossbridge in (a) relaxed and (b) rigor muscle.

these results are in conflict with X-ray studies of relaxed muscle, which have been interpreted as showing that the heads project outwards from the thick filament backbone in a regular fashion.

With the relaxed and rigor states characterized in outline, we can return to the question of what the X-ray reflections tell us about longitudinal movements of crossbridges during contraction. As in rigor, isometrically contracting muscles practically lose the 42.9 nm layer line while the intensity of 14.3 nm meridional may change slightly. Thus the crossbridges move from their helically ordered state about the thick filament backbone. However, in contrast to rigor, there is no detectable increase in the 5.1, 5.9 and 37 nm actin-based reflections. There are, however, intensity changes in the layer lines far from the meridian believed to arise from tropomyosin movements (Chapter 7). The simplest interpretation is that a significant number of crossbridges are attached, based on the 1.0 and 1.1 equatorial reflections, but that their angular distribution is wide. Spectroscopic studies confirm that the head orientation is considerably randomized in contracting fibres compared with those in rigor [34]. Longitudinal movements of crossbridges are therefore observed in contracting muscle but it remains to be shown that these occur when the crossbridges are actually attached.

In our trial of the crossbridge the most serious charge we can prove, using X-ray diffraction evidence alone, is one of loitering with intent.

6.4 Mechanical transients

The independent force generator model implies that crossbridges behave as members of a tug-of-war team. We can learn about the opposition by examining their response to rapid length and tension changes. In a game of tug-of-war if we suddenly release the rope it will move rapidly as the opposing team fall backwards, then it will come to a halt as they pick themselves up and finally it will be withdrawn at a steady velocity as the team run backwards (a velocity transient). Alternatively, if we release the rope by only a small amount the opposing force will drop instantaneously as the team are caught off-guard, but partially recover as they take up the slack by bending elbows and knees. Somewhat later the original force will be restored as they reoptimize their foot holding (a tension transient). Without seeing the opposition we could determine the effective stride and reaction time of an individual but we could not tell whether he used his arms, torso or legs to take up the slack.

Isometrically contracting muscle yields a similar response to such perturbations, although it is difficult to be sure that the applied step is rapid enough to take the crossbridge by surprise. Mechanical devices (Fig. 2.3) have a high inertia and are not suitable for these studies. Fig. 6.10 illustrates the basic layout of a system capable of measuring the steady state and transient behaviour of single fibres just a few millimetres long.

A tension transducer converts the small distortion experienced by its

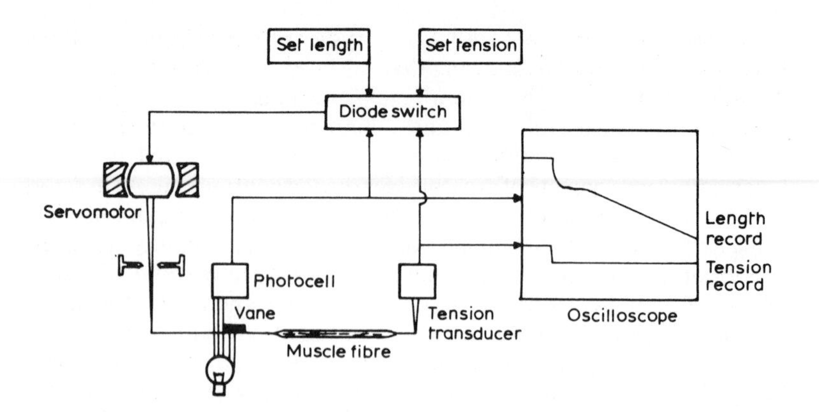

Fig. 6.10 Schematic diagram of system for recording the length and tension of a muscle fibre.

stylus into an electrical signal. Such a device may rely on the change in its resistance or capacitance or take the form of a valve with a movable anode. Either way the physical movement of the stylus must be negligible relative to the desired length parameter imposed on the muscle. The length itself can be controlled by a modified moving-coil galvanometer which acts as a servomotor. Transducing systems for monitoring the length take on many forms. Frequently they are based on the movement of a vane between a light source and a photocell, thereby keeping the inertia very low. The signals from the length and tension transducers are amplified and recorded on an oscilloscope. These signals are also fed back via an electronic switching mechanism to the servomotor allowing the muscle to be kept at constant length or tension. The response time and stability depend on the components and damping circuits used, but all systems have a limit which is a compromise. Good transducers can respond within just under a millisecond.

An important factor in any apparatus is the mode of attachment of the specimen. Various methods involving threads, clamps or glue are used, but all are likely to introduce additional compliance or local damage. Even if satisfactory anchorage is achieved via the tendon, the overall response of the muscle may be influenced by the non-uniform extension of its ends. A further degree of sophistication to the apparatus shown in Fig. 6.10 has been to incorporate a spot-follower [29]. In this device the fluorescent beams from a two-channel cathode ray tube monitor the separation between two specks of gold leaf which are stuck on the central portion of the muscle. By a negative feedback circuit, the servomotor continuously controls the distance between the two gold specks to that desired and therefore compensates for any compliance at the ends of the muscle and in its attachments.

The response of frog muscle to rapid step tension and length changes comprise four phases (Fig. 6.11) which have been interpreted by A.F.

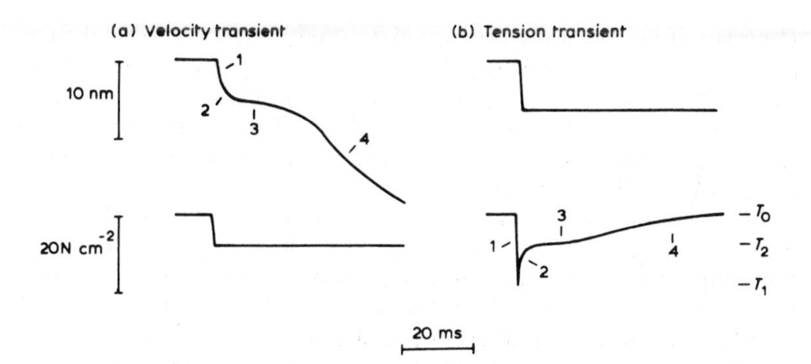

Fig. 6.11 The response of a frog skeletal muscle in isometric tetanic contraction at 2.5°C to (a) a rapid tension change, and (b) a rapid length change. The upper traces are the length records and the lower traces are the tension records as a function of time. The four phases of the transients are described in the text. (After A.F. Huxley [27].)

Huxley and Simmons [35] as follows. Phase 1 occurs concomitantly with the applied step. In a tension transient the instantaneous tension T_1 depends linearly on the size and direction of the applied step, indicating that it involves a Hookean elastic element. Phase 2 represents the rearrangement of attached crossbridges. This leads to a partial recovery of the tension to a value, T_2, which is a non-linear function of the step length (Fig. 6.12). Phase 3 is a plateau caused by detachment of strained crossbridges and reattachment at more optimally placed sites. In phase 4 reattachment continues to restore the original isometric tension T_0, or generate a steady isotonic contraction.

The amplitude of the transient phase (T_1 and T_2) becomes scaled down at long sarcomere lengths (reduced filament overlap) in proportion to the isometric tension, indicating that these responses are largely a reflection of the crossbridges and not some other series elasticity. The interpretation that phases 1 and 2 represent rearrangement of attached crossbridges, rather than detachment and reattachment, is less certain.

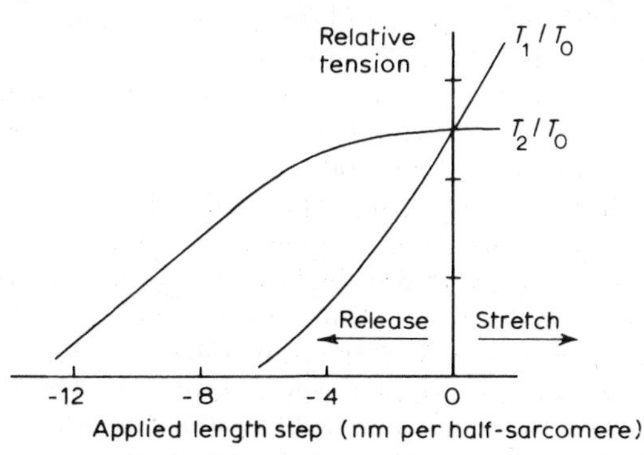

Fig. 6.12 The dependence of the instantaneous tension, T_1, and the early recovery, T_2, observed in a tension transient (see Fig. 6.11b) as a function of the size of the applied stretch or release of a muscle fibre in isometric contraction (T_0). (From A.F. Huxley [27].)

However, if the linear dependence of T_1 on the step change is taken as evidence for passive elasticity in the crossbridge (Fig. 6.12), then its slope (stiffness) provides a measure of the number attached. When a second length step is applied at the end of the phase 2 the stiffness remains the same, suggesting there is no change in the number of attached crossbridges at this point in time.

Assuming the Huxley–Simmons interpretation is correct, Fig. 6.12 provides a measure of the working stroke of the crossbridge. Note that the applied length steps are expressed in terms of nm per half-sarcomere: the effective unit of interaction. The T_2 curve shows that the attached crossbridge can take up about 5 nm of slack to restore the isometric tension. Since, on average, the passive elastic element is effectively stretched by 6 nm from its rest length (the T_1 curve) in the isometric state, the total movement of the crossbridge is about 11 nm. This compares favourably with the assumption that the myosin head (15 nm) moves from an angle of 90° to 45° while attached. However, the calculation ignores the true 3-dimensional relationship between the thick filaments and the actin sites.

Another notable feature of phase 2 is that its rate constant is larger (about 1000 s^{-1}) for greater releases. Huxley and Simmons [35] explained this on the basis of several attached states, in equilibrium under isometric conditions, in which chemical energy is traded for extension of the elastic element. A quick release will perturb this distribution: the greater the release, the less the strain in the elastic element and so the lower the activation energy for redistribution. Note that this model differs from the original Huxley 1957 scheme [15] (Section 6.1) in several respects. There, crossbridges were proposed to attach in a strained state, but this is undesirable on kinetic grounds because it requires that thermal motion overcomes an activation energy of the same magnitude as the overall transduction process. By postulating several bound states the activation energies for individual steps, including attachment, are reduced. However, the original postulate, that the rate of detachment for crossbridges exerting positive tension is slow, remains important for efficient coupling.

As in our tug-of-war game, the domains within the crossbridge responsible for these transients cannot be identified without further information. A number of possibilities have been considered for the location of the elastic element. Many models place a spring-like component in the subfragment 2 region, although the grounds for this may be no more than ease of drawing! In this location the S2 itself must be capable of extending about 11 nm (Fig. 6.12) which is unlikely for a stable coiled-coil structure. A reversible melting of the coil could achieve this shortening but, in unwinding, severe topological problems would arise if the head is to remain attached to actin. The other extreme is to consider that the elasticity is present in the actomyosin bonds. These may undergo an angular distortion and allow the head to achieve 11 nm movement at its distal end. Other possibilities include bending of the myosin head itself or of some elastic domain within each actin monomer. These suggestions are distinguishable, in principle, by monitoring the effect of stretch on head orientation by X-ray diffraction or spectroscopic methods. Extending rigor muscle does not seem to result in a change in attachment angle, although this must be expected if the head was a rigid lever operating on the actomyosin bond [34, 36]. The site of active tension generation is likewise unknown. The simplest conceptual model involves a rotation of the head about the actomyosin bonds, but an angular change could occur elsewhere in the crossbridge.

Recently, the X-ray diffraction pattern of a muscle has been recorded on the time scale of the mechanical transients and provides further evidence for longitudinal motion of crossbridges during contraction [31]. As noted in Section 6.3, the intensity of the 14.3 nm spot arising from the axial spacing of the crossbridges is rather similar in relaxed, rigor and steady state isometric contraction, indicating that the crossbridges select an actin site which least perturbs their regular origins along the thick filament axis. Following a rapid stretch or release the intensity drops markedly, but recovers within about 50 ms. If a quick stretch is applied soon after a quick release, or *vice versa*, some of the intensity is regained immediately. These results imply that, following a length perturbation, many attached crossbridges are distorted longitudinally so that their mass is no longer distributed at regular 14.3 nm intervals. Subsequently, the crossbridges detach and find more favourable actin sites.

Mechanical transients confirm that the crossbridge is a viable entity for active force generation and provide some information on its rate of action. However, definition of the rate constants involves an additional dimension compared with solution kinetics. A complete analysis requires the evaluation of each chemical step as a function of the longitudinal displacement of the actin site, the so-called x dependence.

6.5 Mechanochemical coupling

The shortcomings in our knowledge of crossbridge action prevent a description of the coupling of its chemistry to its mechanics. Instead we are forced to consider models of how crossbridges might work and to check any predictions that might arise. A model in which a myosin head attaches, moves and detaches for each cycle of ATP hydrolysis seems qualitatively consistent with the evidence presented so far. Is this model quantitatively viable and what are its implications?

Consider a frog sartorius muscle contracting at 0°C at optimal efficiency, i.e. lifting a load, $P_0/3$ at a velocity, $V_0/3$ (Figs 2.4 and 2.7). The efficiency is defined by the ratio of the power output (work rate) to ATP utilization rate. From the data in the Appendix, the power from a 1 cm cube of muscle is given by:

$$P_0/3 \times V_0/3 = 6.7 \times 0.00067 \text{ Nm s}^{-1} = 0.044 \text{ J s}^{-1}$$

Taking 1 cm^3 = 1 g of muscle and ΔG for ATP hydrolysis = 60 kJ mol^{-1}

$$\text{ATPase} = 1.5\ \mu\text{mol s}^{-1} = 0.09 \text{ J s}^{-1}$$

Therefore the efficiency of coupling = 0.044 × 100/0.09 = 49%.

We may calculate the stroke size (or at least a limit) of our putative crossbridge cycle in two ways.

1. From the work done (force × distance) per ATP molecule.

   Let a fraction F of the 300 myosin heads in half a thick filament be attached and move, on average, through a distance a nm. Then:
   Force $= P_0/3 = 6.7\ \text{N cm}^{-2} = 1.1 \times 10^{-10}$ N per filament
   $= 3.7 \times 10^{-13}/F$ N per head

   Work done $= 3.7 \times 10^{-13} \times 10^{-9} \times a/F$ J per head per cycle

   Energy available from 1 ATP molecule $= 60 \times 10^3/6 \times 10^{23} = 10^{-19}$ J
   Hence, for an efficiency of 49%, crossbridge stroke:
   $a = 10^{-19} \times 0.49 \times F/3.7 \times 10^{-22} = 130F$ nm

2. From the ATPase rate (6.3 ATP per head per s).

   For a unit of a 1/2 sarcomere $V_0/3 = 833\ \text{nm s}^{-1}$. During 1 ATP cycle let the head move, on average, a nm while attached and free-wheel b nm while detached. Hence $6.3(a+b) = 833\ \text{nm s}^{-1}$.

   For a large population of crossbridges the fraction attached at any one instant, $F = a/(a+b)$
   therefore $a = 833F/6.3 = 130F$ nm

While the actual stroke, a, cannot be determined without knowledge of F, the calculations show that the 11 nm movement, deduced from mechanical studies (Section 6.4), is thermodynamically feasible. Taking 11 nm as an average stroke implies that F is rather small, i.e. at any instant only 8% of the heads are attached. The situation is not that simple, however, because a substantial number of heads may attach transiently without doing any work (i.e. they do not complete a cycle of ATP hydrolysis). Nevertheless the model suggests that the fraction of heads attached should decrease as the contraction velocity increases, because a has an upper limit set by the dimension of the crossbridge and the ATPase rate falls slightly at velocities above $V_0/3$ (i.e. b must increase). Experimental methods for determining F are ambiguous. The 1.1/1.0 equatorial X-ray reflection ratio falls slightly at high velocities of contraction, but this number depends also on the angular distribution of the heads. Furthermore, it cannot distinguish between attached heads and those just near to the actin filament. The instantaneous stiffness (T_1 slope) of isotonically contracting muscle is 20% that of rigor, while the isometric state is about 70% as stiff as rigor [37, 38]. These figures suggest fewer heads are attached in the isotonic state.

Note that the rate constants for processes occurring while the head is attached must exceed $833/a = 75\ \text{s}^{-1}$. How do these data compare with solution kinetics? Here the S1 heads are under no tension and the situation appears crudely analogous to an unloaded contraction. In the Lymn–Taylor model (Fig. 5.3) the V_{max} for the actin-activated S1 ATPase (20 s^{-1} at 20°C, 4.5 s^{-1} at 0°C) was equated with the A.M.D→A.M step, k_6. This appears incompatible with attached transitions having a rate constant $> 75\ s^{-1}$. However, if V_{max} is limited by a refractory state transition in the dissociated head, then the calculated rate constant for muscle ($833/b = 7\ s^{-1}$) appears more reasonable. Unfortunately relating solution studies to an unloaded muscle is more complicated than this. In the latter the heads cannot operate independently in the sense that potential actin sites pass by at a rate determined by the behaviour of the whole sarcomere assembly. The equilibrium and rate constants of the binding steps depend on the relative filament displacement, i.e. on whether or not an actin site is lined up with a myosin site. The elasticity within each crossbridge allows attachment to occur when the sites are not matched perfectly but the equilibrium constant (i.e. free energy) will be changed in proportion to the degree of strain (elastic energy). Likewise the equilibrium between two attached states, which differ in the head angle, will be perturbed from the solution value. To a first approximation the elastic element behaves like a Hookean spring ($\Delta T \propto \Delta L$) and hence the energy required to stretch or compress it is proportional to $\Delta L^2/2$. Simplistically, we can use the solution data to calculate the equilibrium constant (chemical free energy) at zero strain and then modify this number by an amount appropriate to the elastic energy [39]. In this way the energetics can be defined for each step as a function of filament displacement.

Fig. 6.13 shows the Lymn–Taylor model (Fig. 5.3) modified to include an elastic element. It is assumed that the A.M.D state requires the head be at 90° and the A.M state be at 45° to the filament axis. The elasticity is modelled as a spring at the S1–S2 junction, but bear in mind the comments in Section 6.4. The relative longitudinal position of the filaments for the crossbridge under consideration, x, is arbitrarily taken as 0 nm when the A.M.D state is unstrained. Hence, if the head is 15nm long, the A.M state will be unstrained when $x = 11$nm.

In solution at physiological [ADP] and [Pi], the A.M.D→A.M transition lies strongly to the right ($K = 10^6$), but in our model this step can only be achieved by stretching the spring (state b) or allowing the

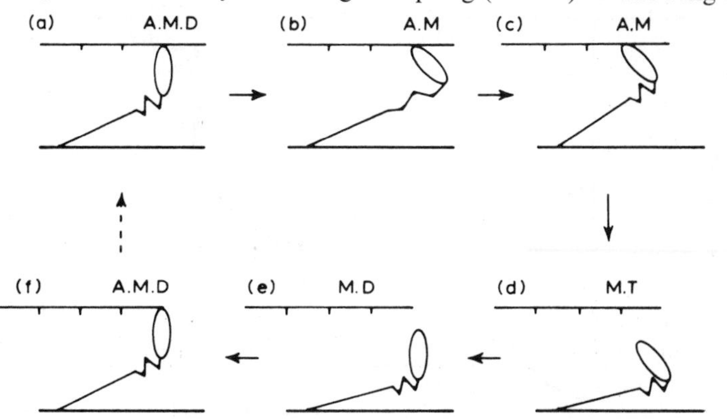

Fig. 6.13 A model for mechanochemical coupling modified from Fig. 5.3 to include an elastic element.

filaments to move (state c). Hence the equilibrium between states a, b and c will depend on x.

On the other hand, the equilibrium between the detached states, d and e, is unaffected by the value of x and therefore the solution value for M.T→ M.D applies directly. In order to keep the comparisons of free energy changes meaningful, each step must be in the form of a unimolecular transition, i.e. binding steps are normalized by taking an effective reactant concentration. The standard free energy so adjusted is termed the basic free energy. Plausible basic free energies of the states depicted in Fig. 6.13 are shown in Fig. 6.14 as a function of x. M.T and M.D are independent of x and separated by only 5.4 kJ mol^{-} because the reaction is readily reversible (Section 5.2, $K_2 = 9$). Attached states show a minimum energy defined by their basic free energies in solution and increase either side as a function of x^2 due to the energy of compressing or stretching the spring. The overall free energy drop for one cycle equals that for ATP hydrolysis (63 kJ mol^{-1}).

Provided that the energy minima of two attached states are displaced with respect to x then work can be extracted from the cycle. The solid line in Fig. 6.14 shows a possible pathway taken by a crossbridge which would give efficient coupling during isotonic contraction. Work is only obtained by releasing the elastic energy of a state as x changes (e.g. b→ c). Free energy released by chemical transitions (shown as vertical drops) is lost as heat. In an isometric contraction x is held constant (although crossbridges throughout the filament will have different x values), hence only vertical transitions can occur. Isometric tension is generated by the crossbridge fluctuating between states a and b, so

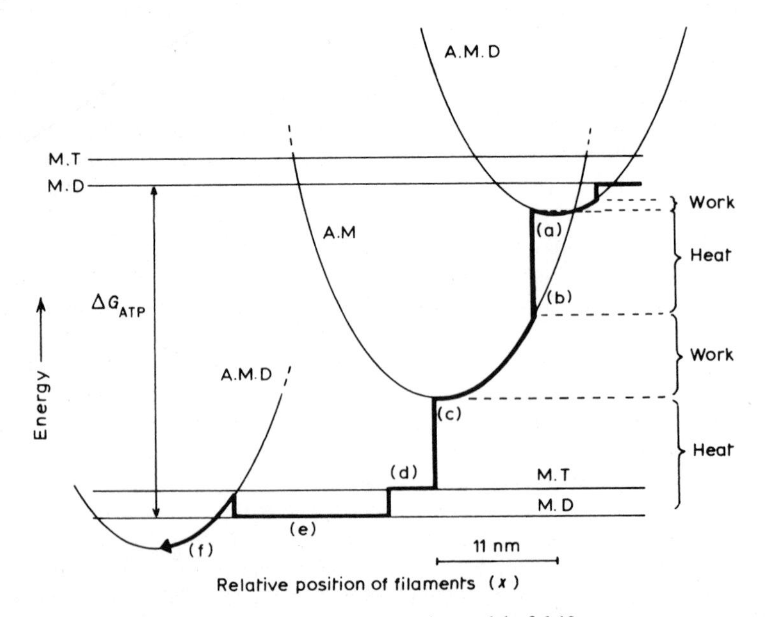

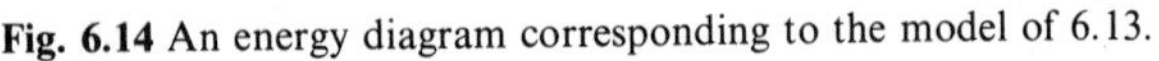
Fig. 6.14 An energy diagram corresponding to the model of 6.13.

trading chemical for mechanical energy, but all the free energy is ultimately released as heat.

Fig. 6.14 is not an answer to the mechanism of energy transduction, but rather, one way of formulating the problem. One crucial assumption is that the energies of the A.M.D and A.M states have a minimum at different x values. The other crucial assumptions for mechanochemical coupling concern the rates of the transitions. Fig. 6.14 defines only the equilibrium constants and therefore we have a choice of forward or reverse rate constants (but not both!). We must assume that a direct b→d transition is prevented, otherwise state c will be bypassed and little tension or work could be obtained. Overall this short circuit is energetically favourable and therefore must be blocked by a high activation energy – in the same way as the spontaneous hydrolysis of ATP is prevented. In the absence of detailed structural information about subfragment 1 and the actin binding site an explanation for this valve-like property cannot be formulated.

Fig. 6.14 does, however, clarify the energy relationships required to give a high efficiency of coupling. The 5.4 kJ mol^{-1} associated with the M.T→ M.D step is unavoidably lost as heat, while the 34.2 kJ mol^{-1} for the A.M.D→ A.M step allows a potential efficiency of coupling of 54%. The change in angle of the detached head probably occurs passively by thermal motion about the S1–S2 hinge, but the properties (conformation?) of the head must depend on the M.T→ M.D step in order for it to bind optimally at different angles. We may explore this proposition experimentally. High ADP concentrations might be expected to induce the 45° state to revert to the postulated 90° attached state (i.e. A.M→ A.M.D; Fig. 6.13). While a small effect is seen in this direction, a larger change is induced by a non-hydrolysable ATP analogue, β–γ imido ATP (AMP.PNP). The experiments were performed by examining the effects of these nucleotides on insect flight muscle in rigor. When a muscle held at a fixed length goes into rigor, it generates a high tension as the crossbridges undergo a final cycle of ATP hydrolysis. However, unlike the isometric state, if a rigor muscle is released by 1% of its length the tension drops to zero and stays there. Addition of ADP or AMP.PNP causes a rigor muscle under tension to 'relax', but if the muscle is then stretched a high tension is re-established, i.e. the muscle has the same stiffness indicating that the heads remain attached. It appears that in the presence of nucleotide, the heads swing back from the A.M state (c) towards the A.M.D state (a), so lengthening the muscle. Electron micrographs and X-ray diffraction confirm this trend in the attachment angle [40]. These findings are perhaps the best evidence available that indicates crossbridges work in a mechanical way under the direct influence of steps in the ATPase cycle. We may conclude that the Lymn–Taylor scheme provides an energetically self-consistent, but incomplete explanation of mechanochemical coupling. There are many potential solutions (i.e. sets of rate constants) for Fig. 6.14 and the problem is to select those which best reflect the physiological performance of muscle [39].

Topics for further reading

Tregear, R.T. and Marston, S.B. (1979), The Crossbridge Theory, *Annu. Rev. Physiol.* **41**, 723–736. (A brief review.)

Squire, J. (1981), *The Structural Basis of Muscular Contraction*, Plenum Press, New York. (A comprehensive treatise with a good introduction to diffraction theory.)

White, D.C.S. and Thorson, J. (1973), The Kinetics of Muscle Contraction, Prog. Biophys. Mol. Biol., **27**, 175–255. (A discussion of mechanical and solution kinetics.)

7 Molecular basis of regulation

7.1 Role of calcium

The ability of a glycerinated fibre to contract depends on the free Ca^{2+} ion concentration (Fig. 2.2). That Ca^{2+} is the physiological agent responsible for activation of contraction is confirmed by experiments using Ca^{2+} indicators such as aequorin. Aequorin is a bioluminescent protein from a jellyfish, which emits light on binding Ca^{2+}. Stimulation of an intact muscle fibre, injected with aequorin, causes a glow of light, the time course of which preceeds that of tension generation [41]. From the light intensity the concentration of free Ca^{2+} is estimated to change from about 0.1 to 10 μM, which is comparable to that required to activate demembranated muscle. The MgATPase activity of myofibrils and crude actomyosin preparations show a dependence on Ca^{2+} in a similar concentration range and this measurement provides the biochemist's probe for regulatory mechanisms.

It is well established that the sarcoplasmic reticulum (Fig. 3.5) is responsible for regulating the sarcoplasmic Ca^{2+} concentration in skeletal muscle. When a muscle is blended, fragments of the sarcoplasmic reticulum reseal to form vesicles which can be purified by high-speed centrifugation. These vesicles are capable of actively accumulating Ca^{2+} ions in the presence of ATP. While the protein responsible for this pumping action is comparatively well characterized [42], little is known about the components responsible for Ca^{2+} release. The volume of the muscle cell dedicated to sarcoplasmic reticulum is roughly proportional to the speed of the contraction cycle. In the toadfish, 25% of the swim bladder muscle is devoted to this structure. The organ is capable of contracting 100 times a second, allowing the fish to produce an audible grunt. However, it remains to be proven that the sarcoplasmic reticulum can accumulate Ca^{2+} with sufficient rapidity to account for relaxation directly. Note that the very rapid action of asynchronous insect flight muscles is achieved by an entirely different mechanism (Section 6.1), in which the free Ca^{2+} concentration remains high during the elastic oscillations of the wing.

Studies on the molecular mechanism of regulation were (and still are) hampered by ubiquitous Ca^{2+} contamination, the lability of the regulatory proteins and the hydrolysis of CaATP at non-physiological $Ca^{2+}:Mg^{2+}$ ratios. EGTA is now used to control the low levels of Ca^{2+} required for relaxation, in an analogous fashion to pH buffers:

$$pCa^{2+} = pK_{Ca} + \log([EGTA]/[CaEGTA]) \qquad (7.1)$$

where pCa is the negative logarithm of the free Ca^{2+} concentration. The value of pK_{ca} depends on the conditions, but is about 6.

7.2 Actin-linked regulation

Although the ATPase of crude actomyosin from vertebrate skeletal muscle is Ca^{2+} sensitive, this property is lost on purification. Ebashi showed that the sensitivity was restored by the addition of two proteins: tropomyosin and troponin [5]. These proteins are associated with the thin filament and from a combination of X-ray diffraction studies, electron microscopy and antibody labelling, their structural arrangement has been determined (Fig. 7.1).

Tropomyosin comprises a coiled-coil structure similar to light meromyosin. Each tropomyosin molecule extends across seven actin units but overlaps with neighbouring tropomyosins so forming a continuous, rather flexible structure.

Tropomyosin alone can either activate or inhibit the actin-activated ATPase of myosin, depending on its relative concentration, but the effects are not dependent on Ca^{2+}. Troponin in stoichiometric amounts to tropomyosin does, however, result in a Ca^{2+}-sensitive system. In the absence of Ca^{2+}, actin activation is abolished and the ATPase activity approaches that of myosin alone. In the presence of Ca^{2+} an ATPase activity equal or slightly higher than that of purified actomyosin is observed. Troponin comprises three subunits (designated TnI, TnC and TnT), the roles of which have been determined by partial reconstitution studies. TnI alone inhibits the ATPase. TnC interacts with TnI and relieves this inhibition but in a Ca^{2+}-insensitive way. TnT completes the system and restores full Ca^{2+} sensitivity characteristic of native troponin. TnC contains four Ca^{2+} binding sites, two of which are specific (i.e. Mg^{2+} binding is very weak) and these appear responsible for triggering the response.

A plausible structural mechanism for regulation has been derived from the X-ray pattern of intact muscle and electron micrographs of thin filament preparations. Tropomyosin takes on the long pitch helix of the actin filament, but its effect on the intensity of the layer lines depends on its azimuthal position. On activation of intact muscle the intensity of the second layer line (18.5 nm) increases, suggesting that tropomyosin

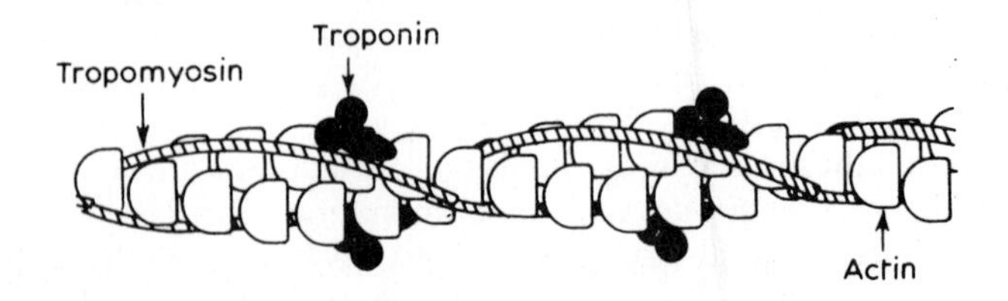

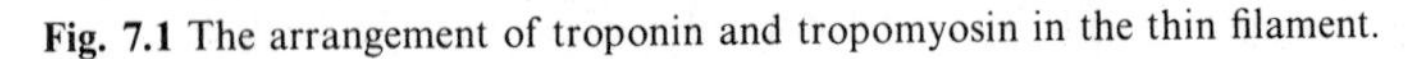
Fig. 7.1 The arrangement of troponin and tropomyosin in the thin filament.

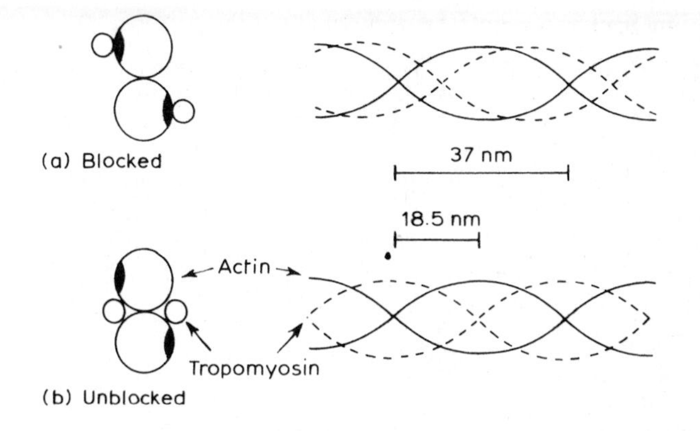

Fig. 7.2 The movement of tropomyosin on activation and its possible role in 'steric blocking'. In cross-section, azimuthal movement of the tropomyosin into the actin groove unblocks the myosin binding site. In a side-on view this appears as a phase shift in the actin and tropomyosin long-pitch helices.

moves towards the centre of the groove. The principle is illustrated in Fig. 7.2. This finding, combined with the knowledge of the position of the myosin binding site identified in decorated thin filaments, has led to a steric blocking model. In the absence of Ca^{2+}, troponin holds tropomyosin towards the edge of the groove, so physically preventing myosin interaction. The binding of Ca^{2+} to troponin induces tropomyosin to move into the groove, and exposes the myosin-binding regions of the actin filament. The details of the model have been revised recently but this hypothesis is retained [20].

Biochemical studies suggest that, in the absence of Ca^{2+}, the regulatory proteins prevent the M.D state (Fig. 5.3) from interacting with actin. At low ionic strength actin binding may ocur but without acceleration of product release. In the absence of ATP the myosin heads have such a high affinity for actin that tropomyosin is pushed into the groove regardless of the Ca^{2+} concentration. At low ATP concentration a limited number of rigor complexes (i.e. the A. M state) will form and render the tropomyosin inoperative, so that neighbouring actin sites will activate ATP hydrolysis even at low Ca^{2+} concentrations. While this effect does not occur at physiological ATP concentrations, it does reveal itself in many *in vitro* experiments.

7.3 Myosin-linked regulation

Myofibrils prepared from molluscan muscles contain very little troponin, at least after washing at low ionic strength, yet their ATPase is Ca^{2+}-sensitive. Moreover, myosin extracted from these muscles exhibits a Ca^{2+}-sensitive ATPase when activated by purified F-actin. The mechanism of regulation was revealed using scallop myosin which, on washing with EDTA, loses one type of light chain and becomes insensitive to Ca^{2+} (i.e. desensitized) [3, 43]. Ca^{2+} sensitivity is restored by adding back this regulatory light chain in the presence of Mg^{2+}. The light chains are located near the S1–S2 hinge and this segment is required for Ca^{2+} sensitivity, suggesting that the mechanism operates on the hinge, rather than by steric blocking of the actin-binding site. The precise location of the specific Ca^{2+} binding sites remains in question.

The discovery of myosin-linked regulation stimulated the search for the role of light chains in other muscle types. All myosins contain two classes of light chains, one of which will bind specifically to desensitized scallop myosin and hence is termed 'regulatory'. Not all regulatory light chains, however, are capable of restoring the Ca^{2+} sensitivity. The other class of light chain has been termed 'essential' because of the difficulty in removing this subunit without loss of ATPase activity. While this nomenclature remains the most general, it is not proven that all 'regulatory' light chains regulate. Moreover, 'essential' light chains might prove to be essential for regulation rather than for ATPase activity! Other ways of naming light chains relate to their electrophoretic mobility, the agents required for their dissociation or their susceptibility to phosphorylation.

Rabbit fast skeletal muscle does not possess a Ca^{2+}-sensitive myosin, at least after extraction, and its 'regulatory' light chain (also known as LC2, DTNB-LC or P-LC) does not impart Ca^{2+} sensitivity to desensitized scallop myosin. However it is clearly homologous in sequence to the scallop regulatory light chain.

Most regulatory light chains are phosphorylated by a specific kinase and dephosphorylated by a specific phosphatase. Those from molluscan muscles appear to be an exception. The physiological significance of phosphorylation remains controversial, but a regulatory role is suspected because an associated protein, calmodulin, imparts Ca^{2+} sensitivity to the kinase. In the case of smooth muscle myosin, phosphorylation of its regulatory light chain has been proposed to be necessary for full actin activation [44]. In non-muscle cells phosphorylation appears to control the assembly of myosin filaments [45]. The rabbit skeletal regulatory light chain becomes phosphorylated during tetanic contraction but the significance remains unknown.

The 'essential' light chain of rabbit skeletal muscle comes in two forms, LC1 (or alkali 1) and LC3 (or alkali 2) which differ only slightly in sequence. The essential light chains are now known not to be obligatory for ATPase activity, although they may modulate actin activation [16]. Within a single muscle fibre, myosin molecules occur which contain two LC1 or two LC2 or one of each light chain type. Isotypes of heavy chains have also been identified and hence conventional myosin preparations contain a complicated mixture of molecules, including some with non-identical heads.

7.4 Multiple regulatory systems

The discovery of different mechanisms of regulation begs the question as to whether a single muscle possesses more than one system. A simple test has been devised to identify the presence of actin- and myosin-linked regulation [43]. Pure rabbit skeletal myosin or pure F-actin is added to a crude actomyosin preparation from the muscle under question and the ATPase is compared in the absence and presence of Ca^{2+}. If the muscle lacks actin-linked regulation the rabbit myosin will interact to give a high ATPase activity regardless of [Ca^{2+}]; while if it lacks myosin-linked regulation a high ATPase activity will be observed on addition of pure actin. Using this criterion, vertebrate skeletal muscles display actin-linked regulation and molluscan muscles myosin-linked regulation. However, many other invertebrates are dually regulated. The result may be ambiguous when phosphorylation is involved because the ATPase can then depend on the activity of endogenous kinases and phosphatases.

The conclusions from such competitive actin and myosin assays, however, remain controversial. It is not clear, for instance, if rabbit skeletal muscle lacks myosin-linked regulation because the proteins are labile, or the ionic conditions of the assay are non-physiological, or its 'regulatory' light chain has become ineffective during the course of evolution. The loss of a labile inhibitory system might explain the rather high ATPase activity of myosin *in vitro* compared with that of resting muscle (Section 4.4). Speculation has also arisen from the discovery that troponin C, myosin light chains, calmodulin and parvalbumin (a soluble Ca^{2+} binding protein of unknown function) possess homologous amino acid sequences, and hence all are implied to be involved in Ca^{2+}-modulated processes [1]. The atomic structure of parvalbumin has been deduced by X-ray crystallography and Ca^{2+} binding domains have been identified in other members of this family by comparison. Some examples are shown in Table 7.1, using the single letter notation for amino acids (see Stryer, p. 16 [46]).

While the homology may not be immediately apparent, a repeating pattern is revealed when the class of amino acid is considered. The first and last eight residues in each domain form an α-helix with an inner face of hydrophobic residues (marked h). The central 13 residues form a potential Ca^{2+} binding loop, with polar or negatively-charged residues at critical positions (X, Y, Z, etc.) which ligate with the metal ion. The residue at −Y donates its backbone carbonyl oxygen and hence its R group is not critical. The affinity and specificity of a particular domain

Table 7.1

Parvalbumin (domain 3)	TKTLMAAG	DKDGDGKIGADE	FSTLVSES
Troponin C (domain 3)	LAECFRIF	DRNADGYIDAEE	LAEIFRAS
Regulatory light chain (domain 1)	FKEATTVI	DQNRDGIIDKED	LRDTFAAM
Essential light chain (domain 1)	FKEAFLLY	DRTGDSKITLSQ	VGDVLRAL
Critical residues	h hh h	X Y Z -Y-X -Z	h hh h

for Ca^{2+} depends on the precise sequence and on pairing with other domains and quaternary interaction. Troponin C contains two Ca^{2+} specific sites (domains 3 and 4) and two Ca^{2+}/Mg^{2+} sites (domains 1 and 2). Myosin light chains also comprise four domains, but some or all of these contain non-conservative substitutions or deletions. The only competent site (domain 1) of the rabbit skeletal regulatory light chain is non-specific and is probably occupied by Mg^{2+} under physiological conditions. A similar domain is found in the scallop regulatory light chain: the Ca^{2+} specific site involved in myosin-linked regulation being located elsewhere. Hence although all these proteins are likely to contain similar structural domains, their function cannot be defined on this basis alone.

To avoid the potential problem of protein lability, intact skeletal muscle has been investigated for signs of myosin-linked regulation. It may be envisaged that, on activation, the release of myosin heads from the thick filament backbone is triggered directly by Ca^{2+}. However, no intensity changes in the 1.0 equatorial or 42.9 nm layer line X-ray reflections are seen when a skeletal muscle, stretched beyond filament overlap (Fig. 6.2), is stimulated [30]. This suggests that the net crossbridge movement associated with activation of muscle at normal sarcomere lengths is dependent on the exposure of actin sites. Whether a myosin-linked regulatory system first 'unlocks' the crossbridge from the thick filament surface remains in question, but it is not required to stimulate the radial movement of the crossbridge.

The presence of multiple regulatory systems within a single muscle fibre would increase the degree of control obtainable from a small change in Ca^{2+} concentration. To what degree this ploy is used by different muscles remains in question. Studies on skinned fibres support the conclusions based on *in vitro* ATPase assays, in that the predominant mode of control in vertebrate skeletal and molluscan adductor muscles is actin-linked and myosin-linked respectively [47].

Topics for further reading

Kendrick-Jones J. and Scholey, J.M. (1981), *J. Muscle Res. Cell Motil.*, **2**, 347–372. (A review of myosin-linked regulation.)

Perry, S.V. (1979), Biochem. Soc. Trans. Lond., **7**, 593–617. (A review lecture on the regulation of striated muscle.)

Squire, J. (1981), *Nature*, **291**, 614–615. (A commentary on the 'steric blocking' model.)

8 Problems and prospects

In the preceding chapters the molecular basis of muscle contraction has been viewed from several standpoints. No one experimental approach has established the mechanism, but together they suggest that force is generated by the mechanical action of crossbridges which act repeatedly to cause filament sliding. What are the problems and prospects of this crossbridge model? Critics continue to question points of principle and detail. Although it has been proved that the crossbridges are the sites of ATPase activity, could they not set up a force which operates over a distance? This hypothesis would dismiss the specific binding between actin and myosin, observed in dilute solution or rigor muscle, as an artefact. While this may be so, it seems more than coincidental that actin and myosin heads undergo cyclic attachment, in solution, during ATPase activity. The nature of this attachment remains unknown, but in general interactions between proteins involve apolar (hydrophobic and van der Waals forces) and polar forces (ionic and hydrogen bonds). If we accept that crossbridges attach to the thin filaments during contraction, simple mensuration tells us that they must do so repeatedly to achieve a net shortening of 1 μm per half-sarcomere. What is the nature of the crossbridge movement while attached? It may involve a local melting, bending or angle change of the crossbridge itself or a domain to which it is attached. The concept of a rigid crossbridge bending through a 45° angle at the actin interface has arisen by inference from the rigor state, but its widespread use in formulating models of contraction should not lead to its blind acceptance. Direct evidence for this mechanism has remained elusive. However, it does accommodate many observations and provides a starting-point for further investigations. To this end the following questions are pertinent.

1. What is the supramolecular structure of the filaments?
2. What are the atomic structures of the contractile proteins?
3. How do the ATPase intermediates correlate with the crossbridge states?
4. What is the nature of the relaxed state – how many regulatory and modulatory systems are involved?
5. How do the filaments assemble during development and how are they replaced continually in mature muscle?

Advances in our understanding are dependent on improvements in individual techniques, the simultaneous application of different techniques, the choice of specimen and the preparative steps employed.

Electron microscopy continues to play a central role in elucidating filament structure and crossbridge disposition. One aim of these studies is to determine the myosin-eye view of the actin site, which will aid theoretical calculations of the 'x-dependence' of the crossbridge steps. It remains controversial as to whether individual myosin heads bifurcate at the S1–S2 junction and attach to different actin filaments, or whether each pair of heads bind to adjacent actin monomers in the same filament. Both interactions can occur *in vitro*. Likewise, it is not known if the two heads operate independently or cooperatively in the manner of climbing a rope.

Relating the detailed information from solution studies with events in intact muscle has been greatly aided by the use of myofibrils and demembranated fibres. As we have seen, the kinetics of acto-subfragment 1 in solution cannot provide a complete description of events in the myofibrillar lattice. Kinetic studies of myofibrils have, until now, yielded limited information because they normally supercontract in the assay mixture. However, by choosing an ATP concentration less than that of the myosin heads, single turnover kinetics can be analysed for a myofibril under isometric conditions – the excess myosin heads remain attached in a rigor state and prevent filament sliding [48].

Demembranated fibres offer greater scope for chemical studies at controlled length and tension, but here the slow diffusion prevents use of rapid-mixing methods. Recently a high concentration of ATP has been generated rapidly *in situ* by the photolysis of caged-ATP, an inactive precursor which contains a photolabile group attached to the γ-phosphate moiety [49].

Progress has also been reported in solution studies where the binding between actin and myosin heads is normally bimolecular and the pseudo first order rates attainable are probably much slower than the first order rates for optimally placed sites *in vivo*. When subfragment 1 is cross-linked to actin it retains a very high ATPase activity at low total protein concentrations [50]. Hence, it appears to mimic events which occur in the myofibrillar lattice.

Application of different techniques to the same preparation is vital to aid interpretation and to expose artefacts. X-ray diffraction is being used to monitor loss of order during the preparative steps of electron microscopy. Combined use of X-ray diffraction and spectroscopic probes is also required to define the perturbations induced by the glycerination and labelling. X-ray diffraction is increasingly used as a tool to study crossbridge kinetics and here advances depend not only on the refinement of equipment but also on the choice of specimen. Better resolution can be achieved by studying a preparation with a slow crossbridge cycling time. The contractile and catch states of molluscan muscle has been exploited to this end. Alternatively, the steps in the cycle might be slowed down by working at sub-zero temperatures in the presence of an antifreeze.

The regulatory systems of muscle contraction offer a wide field of exploration. It is becoming clear that we must consider more than one state of relaxation and contraction. Restimulation of a muscle soon after a tetanic contraction produces a larger twitch tension than that produced by the muscle after it has been relaxed for some time. This 'memory' might reflect residual subthreshold free Ca^{2+} levels, Ca^{2+} bound to slowly exchanging sites, or phosphorylation. Solution studies have yet to mimic the ATPase activities of intact muscle, which may in part be due to damaged or missing regulatory components.

The ultimate test of our knowledge is to reassemble the contractile system from its components, but here we have advanced little beyond the 1940s actomyosin thread. However, much information is to be gleaned from the natural growth and development of muscle. Although striated muscle appears highly regular, proteins in the filaments are renewed on a time scale of days, without interrupting its function. Indeed, its physiological properties can be gradually transformed by substituting new protein isotypes under the influence of nerve–muscle interaction on the genetic machinery. Clearly, research on muscle contraction will remain multidisciplinary.

Topic for further reading

Huxley, Sir Andrew (1980), *Reflections on Muscle*, Liverpool University Press. (A penetrating analysis of advances and retreats in muscle research.)

Recent Symposia

The Mechanism of Muscle Contraction (1972), Cold Spring Harbor Symposium on Quantitative Biology, vol. **37**.

Insect Flight Muscle (1977) (Tregear, R.T., ed.), Elsevier/North Holland, Amsterdam.

Cross-bridge Mechanism in Muscle Contraction (1979) (Sugi, H. and Pollack, G.H., eds), University Park Press, Baltimore.

Development and Specialization of Skeletal Muscle (1980) (Goldspink, D.F., ed.), Socl. Exp. Biol. Seminar Ser. 7. Cambridge University Press.

Appendix

Some useful numbers concerning vertebrate skeletal muscle (frog sartorius at 0°C)	
Sarcomere length	2.0–3.6 μm (2.5 μm rest length)
Thick filament length	1.6 μm (about 300 myosin molecules)
Thin filament length	1.0 μm (about 380 actin molecules)
Thick filament spacing	45 nm (at rest length)
Thick filament content	6×10^{10} filaments per cm^2 (cross section)
Thick–thin filament spacing	22–30 nm (26 nm at rest length)
Thick–thin filament ratio	1:2
Total actin content	600 nmol g^{-1} muscle (about 600 μM)
Total myosin content	120 nmol g^{-1} muscle (about 240 μM heads)
ATPase rate (isotonic)	1.5 μmol ATP/g muscle/s = 6.3 ATP/myosin head/s
ΔG for ATP hydrolysis	60 kJ $mol^{-1} = 10^{-19}$ J per molecule
Isometric tension P_0	20 N cm^{-2}
Maximum velocity V_0	2 muscle lengths per s = 2.5 μM/s/half-sarcomere
Maximum power	44 mW per g muscle
Elastic modulus (instantaneous stiffness) E	4 kN cm^{-2} (isometric) 25 N cm^{-2} (relaxed)
Myosin subfragment 1 ATPase (in solution)	0.01 s^{-1}
Actin-activated subfragment 1 ATPase (in solution)	4.5 s^{-1}

Based on references 8, 9, 21 and 27 (rates and velocities are 5 to 10 times higher for rabbit psoas muscle at 20°C)

References

[1] Kretsinger, R.H. (1980), *C.R.C. Critical Rev. Biochem.*, **8**, 119–174.
[2] Reedy, M.K. (1967), *Am. Zoologist*, **7**, 465–481.
[3] Sellars, J.R., Chantler, P.D. and Szent-Gyorgyi, A.G. (1980), *J. Mol. Biol.*, **144**, 223–245.
[4] Karn, J., McLachlan, A.D. and Barnet, L. (1982), in *Control of Muscle Development* (Pearson, M. and Epstein, H., eds), Cold Spring Harbor Colloquium.
[5] Ebashi, S. (1980), *Proc. Roy. Soc. Lond.*, Ser B, **207**, 259–286.
[6] Homsher, E. and Kean, C.J. (1978), *Annu. Rev. Physiol.*, **40**, 93–131.
[7] Dawson, M.J., Gadian, D.G. and Wilkie, D. (1977), *J. Physiol.*, **267**, 703–735.
[8] Kushmerick, M.J. and Davies, R.E. (1969), *Proc. Roy. Soc. Lond.*, Ser B, **174**, 315–353.
[9] Kushmerick, M.J. and Paul, R. (1976), *J. Physiol.*, **254**, 693–709.
[10] Huxley, A.F. and Niedergerke, R.M. (1954), *Nature*, **173**, 971–973.
[11] Huxley, H.E. and Hanson, J. (1954), *Nature*, **173**, 973–976.
[12] Peachey, L.D. (1965), *J. Cell. Biol.*, **25**, 209–231.
[13] Suck, D., Kabsch, W. and Mannherz, H.G. (1981), *Proc. Nat. Acad. Sci. USA*, **78**, 4319–4323.
[14] Elliott, A. and Offer, G. (1978), *J. Mol. Biol.*, **123**, 505–519.
[15] Weeds, A.G. and Pope, B. (1977), *J. Mol. Biol.*, **111**, 129–157.
[16] Wagner, P.D. and Giniger, E. (1981), *Nature*, **292**, 560–562.
[17] Wells, J.A., Sheldon, M. and Yount, R.G. (1980), *J. Biol. Chem.*, **255**, 1598–1602.
[18] Mendelson, R.A., Morales, M.F. and Botts, J.B. (1973), *Biochemistry*, **12**, 2250–2255.
[19] Thomas, D.D., Ishiwata, S., Seidel, J.C. and Gergely, J. (1980), *Biophys. J.*, **32**, 873–890.
[20] Taylor, K.A. and Amos, L.A. (1981), *J. Mol. Biol.*, **147**, 297–324.
[21] Ferenczi, M.A., Homsher, E., Simmons, R.M. and Trentham, D.R. (1978), *Biochem. J.*, **171**, 165–175.
[22] Taylor, E.W. (1979), *C.R.C. Critical Rev. Biochem.*, **6**, 103–164.
[23] Bagshaw, C.R., Trentham, D.R., Wolcott, R.G. and Boyer, P.D. (1975), *Proc. Nat. Acad. Sci. USA*, **72**, 2592–2596.
[24] Webb, M.R. and Trentham, D.R. (1980), *J. Biol. Chem.*, **255**, 8629–8632.
[25] Lymn, R.W. and Taylor, E.W. (1971), *Biochemistry*, **10**, 4617–4624.
[26] Stein, L.A., Schwarz, R.P., Chock, P.B. and Eisenberg, E. (1979), *Biochemistry*, **18**, 3895–3909.
[27] Huxley, A.F. (1974), *J. Physiol.*, **243**, 1–43.
[28] Huxley, A.F. (1957), *Prog. Biophys.*, **7**, 255–318.
[29] Gordon, A.M., Huxley, A.F. and Julian, F.J. (1966), *J. Physiol.*, **184**, 170–192.

[30] Haselgrove, J.C. and Rodger, C.D. (1980), *J. Muscle Res. Cell Motil.*, **1**, 371–390.
[31] Huxley, H.E., Simmons, R.M., Faruqi, A.R., Kress, M., Bordas, J. and Koch, M.H.J. (1981), *Proc. Nat. Acad. Sci. USA*, **78**, 2297–2301.
[32] Maw, M.C. and Rowe, A.J. (1980), *Nature*, **286**, 412–414.
[33] Offer, G. (1974), in *Companion to Biochemistry*, vol. 1 (Bull, A.T., Lagnado, J.R., Thomas, J.O. and Tipton, K.F., eds), Longman.
[34] Cooke, R. (1981), *Nature*, **294**, 570–571.
[35] Huxley, A.F. and Simmons, R.M. (1971), *Nature*, **233**, 533–538.
[36] Naylor, G.R.S. and Podolsky, R.J. (1981), *Proc. Nat. Acad. Sci. USA*, **78**, 5559–5563.
[37] Goldman, Y.E. and Simmons, R.M. (1977), *J. Physiol.*, **269**, 55–57.
[38] Julian, F. and Morgan, D.L. (1981), *J. Physiol.*, **319**, 193–203.
[39] Eisenberg, E. and Hill, T.L. (1978), *Prog. Biophys. Molec. Biol.*, **33**, 55–82.
[40] Marston, S.B., Rodger, C.D. and Tregear, R.T. (1976), *J. Mol. Biol.*, **104**, 263–276.
[41] Blinks, J.R., Rüdel, R. and Taylor, S.R. (1978), *J. Physiol.*, **277**, 291–323.
[42] MacLennan, D.H. and Holland, P.C. (1975), *Annu. Rev. Biophys. Bioeng.*, **4**, 377–404.
[43] Lehman, W. and Szent-Györgyi, A.G. (1975), *J. Gen. Physiol.*, **66**, 1–30.
[44] Hartshorne, D.J. and Siemankowski, R.E. (1981), *Annu. Rev. Physiol.*, **43**, 519–530.
[45] Scholey, J.M., Taylor, K.A. and Kendrick-Jones, J. (1981), *Biochemie*, **63**, 255–271.
[46] Stryer, L. (1981), *Biochemistry*, Freeman.
[47] Kerrick, W.G.L., Hoar, P.E., Cassidy, P.S., Bolles, L. and Malencik, D.A. (1981), *J. Gen. Physiol.*, **77**, 177–190.
[48] Sleep, J.A. (1981), *Biochemistry*, **20**, 5043–5051.
[49) Hibberd, M.J., Goldman, Y.E., McCray, J.A. and Trentham, D.R. (1981), *Biophys. J.* **33**, 32a.
[50] Mornet, D., Bertrand, R., Pantel, P., Audemard, E. and Kassab, R. (1981), *Nature*, **292**, 301–306.

Index